Lecture Notes
in Business Information Processing   216

More information about this series at http://www.springer.com/series/7911

Boris Delibašić · Jorge E. Hernández
Jason Papathanasiou · Fátima Dargam
Pascale Zaraté · Rita Ribeiro
Shaofeng Liu · Isabelle Linden (Eds.)

Decision Support Systems V – Big Data Analytics for Decision Making

First International Conference, ICDSST 2015
Belgrade, Serbia, May 27–29, 2015
Proceedings

 Springer

Editors

Boris Delibašić
University of Belgrade
Belgrade
Serbia

Jorge E. Hernández
University of Liverpool
Liverpool
UK

Jason Papathanasiou
University of Macedonia
Thessaloniki
Greece

Fátima Dargam
SimTech Simulation Technology
Graz
Austria

Pascale Zaraté
Toulouse 1 Capitole University
Toulouse
France

Rita Ribeiro
UNINOVA – CA3
Lisbon
Portugal

Shaofeng Liu
University of Plymouth
Plymouth
UK

Isabelle Linden
University of Namur
Namur
Belgium

ISSN 1865-1348   ISSN 1865-1356 (electronic)
Lecture Notes in Business Information Processing
ISBN 978-3-319-18532-3   ISBN 978-3-319-18533-0 (eBook)
DOI 10.1007/978-3-319-18533-0

Library of Congress Control Number: 2015937540

Springer Cham Heidelberg New York Dordrecht London

Printed on acid-free paper

Springer International Publishing AG Switzerland is part of Springer Science+Business Media
(www.springer.com)

EURO Working Group on Decision Support Systems

The EWG-DSS is a Working Group on Decision Support Systems within EURO, the Association of the European Operational Research Societies. The main purpose of the EWG-DSS is to establish a platform for encouraging state-of-the-art high-quality research and collaboration work within the DSS community. Other aims of the EWG-DSS are to:

- Encourage the exchange of information among practitioners, end-users, and researchers in the area of Decision Systems.
- Enforce the networking among the DSS communities available and facilitate activities that are essential for the start-up of international cooperation research and projects.
- Facilitate professional academic and industrial opportunities for its members.
- Favor the development of innovative models, methods, and tools in the field of Decision Support and related areas.
- Actively promote the interest of the scientific community toward Decision Systems by organizing dedicated workshops, seminars, mini-conferences, and conference streams in major conferences, as well as by editing special and contributed issues in relevant scientific journals.

The EWG-DSS was founded with 24 members, during the EURO Summer Institute on DSS that took place at Madeira, Portugal, in May 1989, organized by two well-known academics of the OR Community: Jean-Pierre Brans and José Paixão. The EWG-DSS group has substantially grown along the years. Currently, we count over 250 registered multinational members.

Through the years, much collaboration among the group members generated valuable contributions to the DSS field, which resulted in many journal publications. Since its creation, the EWG-DSS has held annual Meetings in various European countries, and has taken active part in the EURO Conferences on decision-making related subjects. Starting from 2015, the EWG-DSS is establishing a new annually organized conference, namely the International Conference on Decision Support System Technology (ICDSST).

The EWG-DSS Coordination Board is composed by: Pascale Zaraté (France), Fátima Dargam (Austria), Rita Ribeiro (Portugal), Shaofeng Liu (UK), Isabelle Linden (Belgium), Jorge E. Hernández (UK/Chile), Jason Papathanasiou (Greece), and Boris Delibašić (Serbia).

Preface

This fifth edition of the EWG-DSS Decision Support Systems LNBIP series represents a selection of reviewed and revised papers of the 1st International Conference on Decision Support System Technology (ICDSST 2015) held in Belgrade, Serbia, during May 27–29, 2015, with the main topic "Big Data Analytics for Decision-Making." This event was organized by the EURO (Association of European Operational Research Societies) Working Group on Decision Support Systems (EWG-DSS).

The EWG-DSS series of International Conferences on Decision Support Systems Technology (ICDSST), starting with the ICDSST 2015 in Belgrade, were planned to consolidate the tradition of annual research events organized by the EWG-DSS group, as well as to offer the European and International DSS Communities, including the academic and the industrial sectors, opportunities to present state-of-the-art DSS developments and to discuss the current challenges that surround Decision-Making processes, focusing on realistic but innovative solutions; as well as on potential new business opportunities.

The selected ICDSST 2015 contributions' scientific areas were as follows: Big data challenges; Big data algorithms (state-of-the-art and available systems); Big data analytics approaches for solving societal decision-making issues; Big data visualization to support decision analysis; Social-networks analysis for decision making; Group and collaborative decision making; Multi-Attribute decision making; Decision-making integrated solutions within open data cross-platforms; Knowledge management and resource discovery for DM; Innovative decision-making methods, Technologies, and real industry applications; among others. This wide and rich variety of themes allowed us, in the first place, to present a summary about some solutions regarding the implementation of decision-making process in a high variety of domains and, in the second place, to highlight their main trends' and research evolution. Moreover, this EWG-DSS LNBIP Springer volume edition has considered contributions selected from a triple-blind paper evaluation method, maintaining this way its traditional high-quality profile. Each selected paper was reviewed by at least three internationally known experts from the ICDSST 2015 Program Committee and external invited reviewers. Therefore, through its rigorous two-stage based triple round review, 8 out of 26 submissions, which correspond to a 31 % acceptance rate, were selected in order to be considered in this 5th EWG-DSS Springer LNBIP Edition.

It is also our pleasure to have Daniel Power's contribution to this edition with the introduction chapter as an invited paper, considering his recognized engagement within the DSS international research community during the last 20 years.

In this context, the selected papers are representative of the current and relevant research activities in the area of decision support systems such as decision analysis for enterprise systems and nonhierarchical networks, integrated solutions for decision support and knowledge management in distributed environments, decision support systems evaluations and analysis through social networks, e-learning context, and their application to real environments.

This EWG-DSS Springer LNBIP Edition includes the contributions described in the sequel.

"'Big Data' Decision Making Use Cases," authored by Daniel Power; "The Roles of Big Data in the Decision-support Process: An Empirical Investigation," authored by Thiago Poleto, Victor Carvalho, and Ana Paula Costa; "Cloud Enabled Big Data Business Platform for Logistics Services: A Research and Development Agenda," authored by Irina Neaga, Shaofeng Liu, Lai Xu, Huilan Chen, and Yuqiuge Hao; "Making Sense of Governmental Activities over Social Media: A Data-driven Approach," authored by Brunella Caroleo, Andrea Tosatto, and Michele Osella; "Data-mining and Expert Models for Predicting Injury Risk in Ski Resorts," authored by Marko Bohanec and Boris Delibašić; "The Effects of Performance Ratios in Predicting Corporate Bankruptcy: The Italian Case," authored by Francesca di Donato and Luciano Nieddu; "A Tangible Collaborative Decision Support System for Various Variants of the Vehicle Routing Problem," authored by Nikolaos Ploskas, Ioannis Athanasiadis, Jason Papathanasiou, and Nikolaos Samaras; "Decision Support Model for Participatory Management of Water Resource," authored by Annielli Cunha and Danielle Morais; "Modeling Interactions Among Criteria in MCDM Methods: A Review," authored by Ksenija Mandić, Vjekoslav Bobar, and Boris Delibašić.

We would like to take this opportunity to express our deepest gratitude to all authors of the submitted papers for being considered in this reviewing process. All papers were of extremely high quality, hence it was a hard task to select the best eight. Therefore, the EWG-DSS Coordination Board Members would also like to express our gratitude and acknowledgment to our colleagues in the Reviewing Team, who voluntarily contributed to the selection and quality of the included papers in this volume, through their rigorous standards and improvement recommendations. The reviewers of this volume were: Alan Pearman, Alejandro Fernandez, Alex Duffy, Ana Paula Cabral, Ana Respício, Antonio Rodrigues, Boris Delibašić, Daniel Power, Dobrila Petrović, Dragana Bečejski-Vujaklija, Dragana Makajić-Nikolić, Evangelos Grigoroudis, Fátima Dargam, Francesca Toni, Francisco Antunes, Gregory Kersten, Irène Abi-Zeid, Isabelle Linden, Jason Papathanasiou, Jorge Freire de Sousa, José Maria Moreno Jimenez, Kathrin Kirchner, Kostas Stathis, Marcos Borges, Maria Franca Norese, Marko Bohanec, Miloš Jovanović, Mohamed Ghalwash, Nikolaos Matsatsinis, Nikos Ploskas, Pascale Zaraté, Pavlos Delias, Priscila Lima, Rita Ribeiro, Rudolf Vetschera, Sean Eom, Shaofeng Liu, Tina Comes, Vladan Radosavljević, and Xiaojun Wang.

Finally, we believe that this EWG-DSS Springer LNBIP Volume has made a high-quality selection of well-balanced and interesting research papers addressing the conference main theme.

May 2015

Boris Delibašić
Jorge E. Hernández
Jason Papathanasiou
Fátima Dargam
Pascale Zaraté
Rita Ribeiro
Shaofeng Liu
Isabelle Linden

Organization

Conference Chairs

Boris Delibašić	University of Belgrade, Serbia
Fátima Dargam	SimTech Simulation Technology, Austria
Pascale Zaraté	IRIT/Toulouse 1 Capitole University, France
Rita Ribeiro	UNINOVA – CA3, Portugal
Shaofeng Liu	University of Plymouth, UK
Isabelle Linden	University of Namur, Belgium
Jorge E. Hernández	Universidad de La Frontera, Chile and University of Liverpool, UK
Jason Papathanasiou	University of Macedonia, Greece

Program Committee

Adiel Teixeira de Almeida	Federal University of Pernambuco, Brazil
Alan Pearman	Leeds University Business School, UK
Alexander Brodsky	George Mason University, USA
Alex Duffy	University of Strathclyde, UK
Alexander V. Smirnov	Russian Academy of Sciences, Russia
Alexis Tsoukias	Université Paris-Dauphine, France
Alicia Diaz	Universidad Nacional de La Plata, Argentina
Ana Paula Cabral	Federal University of Pernambuco, Brazil
Ana Respício	University of Lisbon, Portugal
Antonio Rodrigues	University of Lisbon, Portugal
Boris Delibašić	University of Belgrade, Serbia
Carlos Henggeler Antunes	University of Coimbra, Portugal
Cathy Macharis	Vrije University of Brussels, Belgium
Claudia Rebello da Motta	Universidade Federal do Rio de Janeiro, Brazil
Csaba Csaki	University College Cork, Ireland
Daniel Power	University of Northern Iowa and DSS Resources, USA
Dobrila Petrovic	Coventry University, UK
Dragana Bečejski-Vujaklija	University of Belgrade, Serbia
Dragana Makajić-Nikolić	University of Belgrade, Serbia
Evangelos Grigoroudis	Technical University of Crete, Greece
Fátima Dargam	SimTech Simulation Technology/ILTC, Austria
Francesca Toni	Imperial College London, UK
Francisco Antunes	Beira Interior University, Portugal
François Pinet	Cemagref/Irstea, France

Sean Eom	Southeast Missouri State University, USA
Shaofeng Liu	University of Plymouth, UK
Slobodan Vučetić	Temple University, USA
Tina Comes	University of Agder, Norway
Uroš Rajkovič	University of Maribor, Slovenia
Vladan Radosavljević	Yahoo Labs, USA
Xiaojun Wang	University of Bristol, UK
Zenon Michaelides	University of Liverpool Management School, UK

Organizing Committee

EWG-DSS

Pascale Zaraté	IRIT/Toulouse 1 Capitole University, France
Fátima Dargam	SimTech Simulation Technology, Austria
Rita Ribeiro	UNINOVA – CA3, Portugal
Shaofeng Liu	University of Plymouth, UK
Isabelle Linden	University of Namur, Belgium
Jorge E. Hernández	Universidad de La Frontera, Chile and University of Liverpool, UK
Jason Papathanasiou	University of Macedonia, Greece
Boris Delibašić	University of Belgrade, Serbia

Local Organizing Team

Boris Delibašić	University of Belgrade, Serbia
Milija Suknović	University of Belgrade, Serbia
Miloš Jovanović	University of Belgrade, Serbia
Sandro Radovanović	University of Belgrade, Serbia

Main Sponsors

Working Group on Decision Support Systems
(http://ewgdss.wordpress.com/)

Association of European Operational Research Societies
(www.euro-online.org)

Institutional Sponsors

**Faculty of Organisational Sciences,
University of Belgrade, Serbia**
(http://www.fon.bg.ac.rs/eng/)

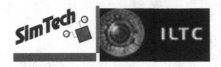

SimTech Simulation Technology, Austria
(http://www.SimTechnology.com)
**ILTC - Instituto de Lógica Filosofia e Teoria
da Ciência, RJ, Brazil** (http://www.iltc.br)

University of Toulouse, France
(http://www.univ-tlse1.fr/)
**IRIT Institut de Research en Informatique
de Toulouse, France** (http://www.irit.fr/)

**UNINOVA - CA3 - Computational
IntelligenceResearch Group**
(www.uninova.pt/ca3/)

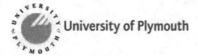

**School of Management, University
of Plymouth, UK**
(http://www.plymouth.ac.uk/)

University of Namur, Belgium
(http://www.unamur.be/)

**Management School, University
of Liverpool, UK**
(http://www.liv.ac.uk/management/)

University of Macedonia, Department of Marketing and Operations Management, Greece (http://www.uom.gr/index.php?newlang=eng)

Universidad de La Frontera, Temuco Chile (www.ufro.cl)

Commercial Sponsors

Arena Simulation Software (https://www.arenasimulation.com)

Lumina Decision Systems (www.lumina.com)

Flexsim Simulation Software (https://www.flexsim.com)

ExtendSim Power Tools for Simulation (http://www.extendsim.com)

Logit Software Serveices & Solution Engineering (http://www.logit-solutions.com/)

In2 Engineering (www.in2.rs)

Contents

'Big Data' Decision Making Use Cases 1
  Daniel J. Power

The Roles of Big Data in the Decision-Support Process: An Empirical
Investigation. 10
  Thiago Poleto, Victor Diogho Heuer de Carvalho,
  and Ana Paula Cabral Seixas Costa

Cloud Enabled Big Data Business Platform for Logistics Services:
A Research and Development Agenda. 22
  Irina Neaga, Shaofeng Liu, Lai Xu, Huilan Chen, and Yuqiuge Hao

Making Sense of Governmental Activities Over Social Media:
A Data-Driven Approach 34
  Brunella Caroleo, Andrea Tosatto, and Michele Osella

Data-Mining and Expert Models for Predicting Injury Risk in Ski Resorts . . . 46
  Marko Bohanec and Boris Delibašić

The Effects of Performance Ratios in Predicting Corporate Bankruptcy:
The Italian Case 61
  Francesca di Donato and Luciano Nieddu

A Tangible Collaborative Decision Support System for Various Variants
of the Vehicle Routing Problem 73
  Nikolaos Ploskas, Ioannis Athanasiadis, Jason Papathanasiou,
  and Nikolaos Samaras

Decision Support Model for Participatory Management of Water Resource. . . 85
  Annielli Cunha and Danielle Morais

Modeling Interactions Among Criteria in MCDM Methods: A Review 98
  Ksenija Mandic, Vjekoslav Bobar, and Boris Delibasic

Author Index 111

Contents

'Big Data' Decision Making Use Cases

Daniel J. Power[✉]

University of Northern Iowa and DSSResources.com,
Cedar Falls, IA, USA
daniel.power@uni.edu

Abstract. New data streams from social media, passive data capture and other sources are creating opportunities to support decision making. Also, data volume, data velocity and data variety continue to increase. Data-driven decision making using these new data streams, often call "big" data, is an important topic for continuing discussion and research. Given the costs of this data it is important to understand "big" data and any decision making use cases. Current use cases demonstrate how new data streams can support some operating decisions. Claims that new data streams can support strategic decision making by senior managers have not been demonstrated. Managers want better data and desire the "right" data at the "right time" and in the "right format" to support targeted decisions. This article explores the challenges of identifying novel use cases relevant to decision making, especially important, strategic long-term decisions. Analyzing "big data" to find a great business plan or to identify the next revolutionary product idea seems however like wishful thinking. Data is useful and we have more of it than ever before and the volume is increasing because data capture and storage is inexpensive. "Big data" and advanced analytics may provide facts for experienced and talented strategic decision makers, but those uses are not clearly defined. At present, the major strategic decision related to "big" data for senior managers is how much time, talent and money to allocate to capturing, storing and analyzing new data streams. Better defined decision making use cases can help senior managers assess the value of new data sources.

Keywords: Data-driven decision making · New data streams · Strategic decision making · Analytics · Computerized decision support · Big data · Use cases

1 Introduction

Managers face both challenges and opportunities from the expanding data streams, e.g., blogs, server log data, web application data, historic transaction data, and internet feeds. The data storage space is rapidly expanding with many new data sources and types. Some observers have called the expanding streams "big data" and "fast data", some characterize the phenomenon in terms of specific dimensions, i.e., high variability with rapidly increasing volume, variety, volatility, and velocity [17]. Supporting decision making and creating new products and services using new data streams is an important topic and more research, discussion and analysis is needed about use cases especially those related to strategic decisions.

© Springer International Publishing Switzerland 2015
B. Delibašić et al. (Eds.): ICDSST 2015, LNBIP 216, pp. 1–9, 2015.
DOI: 10.1007/978-3-319-18533-0_1

Administrative decisions are routine, recurring decisions. Operational or tactical decisions are technical decisions which support execution of strategic decisions. Strategic decisions are choices made by managers that are broad in scope, long term, and risky. These important decisions consider the environment and competitive situation in which a firm or organization operates, its resources and people. Strategic decisions are more complex and unstructured than administrative or operational decisions. Strategic decision makers are the senior managers in an organization who have the responsibility to make long-term, high risk decisions, cf., [13].

Some of the strategic decisions that are most challenging for senior managers include launching a new product, entering a new market, acquiring a competitor and hiring and promoting managerial talent. This article explores using analytics and "big data" to support decision making, especially for strategic decisions. The next section examines the dimensions of the new data streams. Section 3 discusses organization decision making at three levels, operational, management control and strategic. Section 4 explores the challenges of data-driven strategic decision making. The final section draws conclusions about "big data" decision making use cases.

2   New Data Streams

The term "big data" is a colorful, popularized phrase for the new data streams from social media, computing machine logs, and other emerging data sources. Some information technology vendors over promote technology opportunities and that has happened with "big data" and to some extent with the term analytics and data science. "Big data" is entering the Gartner Hype Cycle Trough of Disillusionment [7] for managers and researchers, but new data streams are a reality and decision makers can potentially derive benefits from analyses of data in the expanded data environment.

Figure 1 graphically depicts the expanding multidimensional data space. The big data metaphor is any data at the extreme end of one of three dimensions volume, velocity, and variety. Other dimensions that have been suggested include veracity, viscosity, volatility, and variability. Data variety refers to the increasing number of formats of digital data. Data velocity refers to the increasing speed of data generation, data processing and data use. Data volume refers to the increasing number of bytes of data stored on various media. Data volatility means 'data flows can be highly inconsistent with periodic peaks'. Variable data has a wide spread, is complex and from diverse sources. Variability also means changeable. Metaphorically the data space is an expanding regular pentagon with at least five attributes, cf., [17].

The overall rate of data expansion for all organizations and society in general is increasing rapidly. Some estimates indicate that in 2014 there was 4.4 trillion gigabytes of data and that volume will predicted to expand to 44 trillion bytes by 2020 [20].

One of the major new data streams is from social media. On October 29, 2014, IBM and Twitter [9] announced a partnership to "help transform how businesses and institutions understand their customers, markets and trends – and inform every business decision." Dick Costolo, Twitter CEO is quoted in the press release [9] stating "This important partnership with IBM will change the way business decisions are

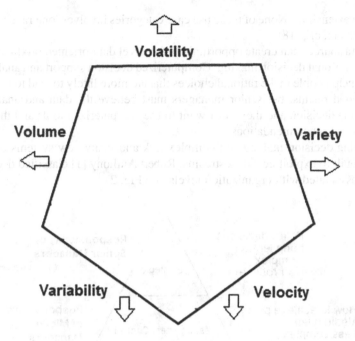

Fig. 1. Expanding multiple dimensions data space

made – from identifying emerging market opportunities to better engaging clients, partners and employees."

Many Information Technology industry analysts and bloggers discuss strategic opportunities from the analysis of new data streams. There are however generally few specifics about how to use the new data streams to gain value. Big data must be analyzed to inform decision making by managers within an organization or by customers of a data and analytics product, but it is not yet clear if or how the expanding data space can support novel decision making, especially strategic decisions. Managers want better data and desire the "right" data at the "right time" and in the "right format" to support targeted decisions. The right data is needed to conduct an appropriate analysis. In general, the "right" data to support decision making is relevant, accurate and timely [14, 16].

3 Organization Decision Making Use Cases

IBM describes five game changing big data use cases, cf., [10]. Datameer [5] describes five similar use cases. The IBM website at http://www-01.ibm.com/software/data/bigdata/use-cases.html explains that "a use case helps you solve a specific business challenge by using patterns or examples of technology solutions."

The five Datameer and IBM big data uses cases are: (1) Customer analytics, (2) Data-driven products and services, including personalization systems, (3) Enterprise Data Warehouse optimization and data warehouse modernization, (4) Operational analytics and monitoring business operations, and (5) Fraud detection, compliance, security/

intelligence extensions. None of these use case categories involves long range, strategic decision making, cf., [18].

New data sources can create opportunities for novel data-oriented products and can improve operational decision making. Computerized decision support and analytics can potentially help people make rational choices that are more likely to lead to goal attainment and good results, but senior managers must believe the data and analytics are relevant to the decision and they must want to use computerized tools and then act on the findings and recommendations.

Supporting decision making is a complex task and many new systems likely will need to be built to exploit new data streams. Robert Anthony [1] classified decisions in categories associated with organization levels (see Fig. 2).

Fig. 2. Three organization decision making general use cases

In general, current applications for the new data sources and analytics are focused on tactical, operational control and operating decisions. For example, middle managers use new data sources and analytics for monitoring product/service quality and operations personnel use results from analysis of new data streams to make day-to-day decisions as part of ongoing operational activities. The goal of analytics is to increase the rationality and speed of these routine, recurring decisions [16].

Many senior executives want operating personnel and lower-level managers to use predictive analytics and "big data" decision support applications. Also, some senior managers want to hire data analysis specialists, data scientists and managers who understand "big data" applications. There is little evidence that senior managers want or think it possible that "big data" and analytics can or will impact long-term strategic decision making.

"Big data" software and hardware salespeople however, seem to think the possibilities for changing companies and the world with "big data" are unbounded. A fundamental shift is supposedly happening in how decisions are made because of "big data".

New data does mean more and better measurement, more record-keeping and more analysis, but there is no guarantee that the data will influence significant or strategic management decisions in a meaningful way. Senior managers must want to use data and analytical decision support if the envisioned data revolution is to actually occur. More data does not mean more relevant or more meaningful and useful data. Also, using data mining and other tools to analyze data does not guarantee results that are meaningful. For example, analyses based upon correlations do not establish causal connections.

Let's examine some "big data" use cases identified by Laskowski [11] and others. A number of sources identify a fast food restaurant chain's use of cameras to determine what to display on the drive-up menu screen as a "big data" decision making use case. Depending on waiting line length, the menu screen will display either items that are quick to prepare or items that may take longer. Perhaps image recognition and inferring waiting line length is a novel data use, but such an application is automating an operating level decision that a person could easily make and then quickly, and manually change or toggle the electronic data display.

Another decision support use case that has been cited is police departments using software from PredPol, the "Predictive Policing Company". The PredPol (http://www.predpol.com/) software uses crime data for a geographic area to assign probabilities for future crime events to regions of space and time and a map software display presents estimated crime risk in a user friendly framework to a law enforcement decision maker like a shift sergeant or lieutenant. Again this use case is for an operational decision making task and the volume, velocity and variety of data used is moderate volume, medium velocity and structured.

Information Management [21] in a slideshow highlights five big data use cases. One of the use cases is summarized with the phrase "connect with outside patterns." The authors explain "Huge sets of governmental or publicly available data help you shape your business plans. The example is however an operational process improvement "one vacation resort drastically cut labor costs by syncing up its scheduling process with information from the National Weather Service." Another cited use case is analyzing big data to help manage disasters and outages. A third example is "big data enables better demographic insight into how products and services are used by your enterprise and your competitors. These use cases lack specifics necessary to develop a model.

Other examples of big data use cases include differential pricing, campaign lead generation, improving the customer experience, marketing mix modeling, predicting customer purchase behavior, and reducing fraud. All of these examples appear to fall in the category of operating and operational control organizational decisions.

Bernard Marr [12], a self-proclaimed Big Data Guru identified nine "high value big data use cases" that he thought were changing the World, include some uses for improved decision making. For example he identified targeting customers, optimizing business processes, and performance optimization. The final decision making use case Marr discusses is High-Frequency Trading (HFT). He explains "big data algorithms are used to make trading decisions. Today, the majority of equity trading now takes place via data algorithms that increasingly take into account signals from social media networks and news websites to make buy and sell decisions in split seconds."

It is generally assumed that managers want to make data-driven decisions, rather than solely intuitive or "gut feel" decisions [15]. Some evidence suggests however that senior managers are primarily encouraging data-driven decision making for operating decisions.

There remain many unanswered questions about big data uses: Are there important, strategic, long-term decision making use cases? Do senior business and government executives want to use "big data" to support their strategic decision making? Are there any examples of using "big data" to support strategic decision makers? What data is the "right data" for supporting a corporate acquisition decision? What data is the "right data" for creating and deciding to launch a new product? How many different types of decisions are supported with big data analytics?

4   Data-Driven Strategic Decision Making

Data analytic vendors assert the best decisions are made based on data and that more data needs to be analyzed. A recent Capgemini study [3] finds that 9 out of 10 business leaders "believe data is now the fourth factor of production, as fundamental to business as land, labor and capital." That report concludes "Big Data represents a fundamental shift in business decision making."

Power [15] argued that it is important for managers and Information Technology professionals to understand data-driven decision support systems and how such systems can provide business intelligence and performance monitoring. The literature provides few examples of data-driven decision support for strategic decisions. One might argue that using a model-driven DSS for capital budgeting decisions or analyses of large data sets to identify companies to acquire are examples of data analytics supporting strategic decisions.

In a Deloitte Review article, Guszcza and Richardson [8] state "Today few doubt that, properly planned and executed, data analytic methods enable organizations to make more effective decisions. Anecdotal evidence abounds." They are more skeptical about the necessity of using big data, noting it is false that big data is necessary for analytics to provide big value.

Former Chairman of the Board of Governors of the IBM Academy of Technology, Irving Wladawsky-Berger [23] noted in a guest column in the *Wall Street Journal* that "Decision making has long been a subject of study and given the explosive growth of Big Data over the past decade, it's not surprising that data-driven decision making is one of the most promising applications in the emerging discipline of data science." He explores the use of big data in decision-making and concludes "the use of Big Data and data science to help with strategic decisions is in its early stages and requires quite a bit more research to understand how to use them under different contexts."

Provost and Fawcett [19] define data-driven decision making as "the practice of basing decisions on the analysis of data rather than purely on intuition." They state "The benefits of data-driven decision making have been demonstrate conclusively." They cite a study by Economist Erik Brynjolfsson and his colleagues from MIT and Penn's Wharton School to substantiate the claimed benefits.

Brynjolfsson, Hitt, and Kim [2] did report that effective use of data and analytics correlated with a 5 to 6 percent improvement in productivity, as well as higher profitability and market value. The study used survey data on the business practices and information technology investments of 179 large publicly traded firms collected in 2008. The survey was conducted in conjunction with McKinsey and Company. Their key independent variable, data-driven decision-making (DDD), combined responses from three questions in the survey: (1) the usage of data for the creation of a new product or service, (2) the usage of data for business decision-making in the entire company, and (3) the existence of data for decision-making in the entire company. The study did not assess using analytics with new data streams like social data.

According to Ehrenberg [6], "Greater access to data and the technologies for managing and analyzing data are changing the world." Many bloggers seem confident that "big data" will lead to better health, better teachers and improved education, and better decision-making. Some skepticism about these claims seems justified given the current evidence. Scientists need to conduct more research about the impact of data-driven decision making using analytics and both small and "big" data sets.

5 Conclusions

Senior managers attempt to make rational and thoughtful decisions, but there is little indication that "big data" can be useful in supporting strategic, long-range decisions. The practice of creating decision support is guided by what works and that is perhaps as it should be given the demands placed by managers to produce useful results [14, 16].

Data is often useful to assist in making decisions [15]. Organizations have however more data than can currently be used and the volume will increase because data capture and storage is inexpensive. Today we can easily capture and analyze data in real-time, so we do that. Data in formats that are unstructured like twitter feeds can now be captured and stored and in some situations the data can meaningfully be analyzed. Managers are finding novel uses for data to support decision making and more uses will likely be identified. Strategic uses are much harder to identify. Both corporate-level managers and senior government decision makers know that "wicked", non-routine decisions are still the primary responsibility of smart, well-informed people. One hopes those senior decision makers use all of their capabilities, request and read targeted research analyses and use facts to make the best decisions humanly possible in those unstructured decision situations [14].

Analyzing "big data" to find a great business plan or to identify the next revolutionary product idea seems like wishful thinking. Trying to use "big data" for making strategic decisions reminds me of the story [22] of a small boy who woke up on Christmas morning to find a huge pile of horse manure in the living room by the Christmas tree instead of presents. His parents discover him happily and enthusiastically digging in the manure. They ask "What are you doing son?" The boy exclaims, "With all this manure, there must be a pony here somewhere!"

Developing an infrastructure of people, processes and technology to capture, store and analyze new data streams is costly. The immediate strategic decision that must be made by senior managers is how much to invest in analyzing new data streams. New data streams and analytics can add value and improve operating decisions [4, 16]. Evaluating potential operating decision making use cases can help managers decide if resources should be dedicated to exploiting "big data". Vendors do not currently provide use cases that show how "big data" can help managers make strategic decisions in organizations. Big data is a reality, but the claims of vendors about its uses often seem over stated. More effort needs to be devoted to developing use case diagrams that show how an analytical task is performed using a new data source with one or more of the "big data' characteristics of high volume, high variety, high velocity, high volatility and high variability.

Overall, there should be optimism about finding novel use cases for new data sources. Managers should be concerned however that in some situations the "big data" will be just a huge pile of horse manure with no or limited value. Also, tools to manage "big data" and analytics technologies are improving, but machine leaning and analytics will not replace the critical thinking skills of experienced and talented human strategic decision makers in the near future [16]. At present, the major strategic decision related to "big" data for senior managers is how much time, talent and money to allocate to capturing, storing and analyzing new data streams. Better defined decision making use cases can help senior managers assess the value of new data sources.

Acknowledgments. Some of the content in this article has appeared in columns published in Decision Support News and is stored at DSSResources.com. The comments of the reviewers stimulated a revision and clarification of the positions in this article.

References

1. Anthony, R.N.: Planning and Control Systems: A Framework for Analysis. Harvard University Press, Cambridge (1965)
2. Brynjolfsson, E., Hitt, L.M., Kim, H.H.: Strength in numbers: how does data-driven decisionmaking affect firm performance? Working paper, Social Science Research Network (SSRN), April 2011
3. Capgemini, Inc.: The Deciding Factor: Big Data & Decision Making, 4 June 2012. http://www.capgemini.com/resources/the-deciding-factor-big-data-decision-making
4. Davenport, T., Patil, D.J.: Data scientist: the sexiest job of the 21st century. Harv. Bus. Rev. **90**, 70–76 (2012). http://hbr.org/2012/10/data-scientist-the-sexiest-job-of-the-21st-century
5. Datameer eBook: Top Five High-Impact Use Cases for Big Data Analytics (2014). http://www.datameer.com/pdf/eBook-Top-Five-High-Impact-UseCases-for-Big-Data-Analytics.pdf
6. Ehrenberg, R.: What's the big deal about Big Data? InformationArbitrage.com blog post, 19 January 2012. http://informationarbitrage.com/post/16121669634/whats-the-big-deal-about-big-data
7. Gartner: Research Methodologies: Gartner Hype Cycle (2015). www.gartner.com/technology/research/methodologies/hype-cycle.jsp. Accessed 3 Mar 2015
8. Guszcza, J., Richardson, B.: Two dogmas of big data: understanding the power of analytics for predicting human behavior. Deloitte Rev. (15), 161–175 (2014). http://dupress.com/articles/behavioral-data-driven-decision-making/

9. IBM: Twitter and IBM form global partnership to transform enterprise decisions. Press Release, 29 October 2014. http://dssresources.com/news/4182.php

10. IBM: The top five ways to get started with big data. Thought Leadership White Paper, Document Number: IMW14710USEN, June 2014

11. Laskowski, N.: Ten big data case studies in a nutshell. SearchCIO, October 2013 http://searchcio.techtarget.com/opinion/Ten-big-data-case-studies-in-a-nutshell

12. Marr, B.: 9 Amazing Ways Big Data Is Used Today to Change the World. SmartData Collective, 5 November 2013. http://smartdatacollective.com/Big_Data_Guru/9-amazing-ways-big-data-used-today-change-world. Accessed 5 Mar 2015

13. Planning Glossary. http://planningskills.com/glossary/

14. Power, D.J.: Decision Support Systems: Concepts and Resources for Managers. Greenwood/Quorum Books, Westport (2002)

15. Power, D.: Understanding data-driven decision support systems. Inf. Syst. Manag. **25**(2), 149–154 (2008)

16. Power, D.: Decision Support, Analytics, and Business Intelligence, 2nd edn. Business Expert Press, New York (2013)

17. Power, D.: Using 'Big Data' for analytics and decision support. J. Decis. Syst. **23**(2), 222–228 (2014)

18. Power, D.J.: What are some use cases with expanded data sources? Decis. Support News **15**(26), 21 December 2014. http://dssresources.com/faq/index.php?action=artikel&id=303. Accessed 3 Mar 2015

19. Provost, F., Fawcett, T.: Data science and its relationship to big data and data-driven decision making. Big Data **1**(1) (2013). http://online.liebertpub.com/doi/pdf/10.1089/big.2013.1508

20. Solomon, H.: The amount of data we are creating is out of this world. ITWorld Canada, 15 April 2014. http://www.itworldcanada.com/article/the-amount-of-data-were-creating-is-out-of-this-world/91586

21. Unknown: 5 big data use cases. Information Management. http://www.information-management.com/gallery/5-big-data-use-cases-10023102-1.html

22. Unknown: There must be a pony somewhere. Quote Investigator. http://quoteinvestigator.com/2013/12/13/pony-somewhere/

23. Wladawsky-Berger, I.: Data-driven decision making: promises and limits. Wall Street J., 27 September 2013. http://blogs.wsj.com/cio/2013/09/27/data-driven-decision-making-promises-and-limits/

The Roles of Big Data in the Decision-Support Process: An Empirical Investigation

Thiago Poleto[✉], Victor Diogho Heuer de Carvalho, and Ana Paula Cabral Seixas Costa

Universidade Federal de Pernambuco, Av. Prof. Moraes Rego 1235, Recife,
Pernambuco 50670-901, Brazil
{thiagopoleto,apcabral}@hotmail.com, victorheuer@gmail.com

Abstract. The decision-making process is marked by two kinds of elements: organizational and technical. The organizational elements are those related to companies' day-to-day functioning, where decisions must be made and aligned with the companies' strategy. The technical elements include the toolset used to aid the decision making process such as information systems, data repositories, formal modeling, and analysis of decisions. This work highlights a subset of the elements combined to define an integrated model of decision making using big data, business intelligence, decision support systems, and organizational learning all working together to provide the decision maker with a reliable visualization of the decision-related opportunities. The main objective of this work is to perform a theoretical analysis and discussion about these elements, thus providing an understanding of why and how they work together.

Keywords: Decision support · Decision-making process · Big Data · Business Intelligence (BI) · Decision Support System (DSS) · Organizational learning

1 Introduction

Organizations need to use a structured view of information to improve their decision-making process. To achieve this structured view, they have to collect and store data, perform an analysis, and transform the results into useful and valuable information. To perform these analytical and transformational processes, it is necessary to make use of an appropriate environment composed of a large and generalist repository, a processor core with the appropriate intelligence (Business Intelligence [BI]), and a user-friendly interface.

The repository must be filled with data originating from many different kinds of external and internal data sources. These repositories are the data warehouses (generalists) and data marts (when considering a specific company activity or sector), and most recently, Big Data.

The Big Data concept and its applications have emerged from the increasing volumes of external and internal data from organizations that are differentiated from other databases in four aspects: volume, velocity, variety, and value. Volume considers the data amount, velocity refers to the speediness with which data may be analyzed and processed,

© Springer International Publishing Switzerland 2015
B. Delibašić et al. (Eds.): ICDSST 2015, LNBIP 216, pp. 10–21, 2015.
DOI: 10.1007/978-3-319-18533-0_2

variety describes the different kinds and sources of data that may be structured, and value refers to valuable discoveries hidden in great datasets [1].

Big Data has the potential to aid in identifying opportunities related to decision in the intelligence phase of Simon's [2] model. In some cases, the stored data may be used to aid the decision-making process. In this context, the term "intelligence" refers to knowledge discovery with mining algorithms. In this way, Big Data use can be aligned with the application of Business Intelligence (BI) tools to provide an intelligent aid for organizational processes. The data necessary to obtain the business perceptions must be acquired, filtered, stored, and analyzed after the available data are heterogeneous and in a great volume. The processes of filtering and analysis of the data are very complex, because of that it is necessary the use BI strategies and tools.

The main proposal of the present study is to develop an investigation that describes the roles of Big Data, and BI in the decision-making process, and to provide researchers and practitioners with a clear vision of the challenges and opportunities of applying data storage technologies so that new knowledge can be discovered.

The sequence of this work is as follows. Section 2 provides a background for Big Data and some of its applications. Section 3 introduces the concept of DSS. Section 4 conceptualize BI and presents its organizational and technological components. Section 5 presents a scheme for the integration between Big Data, BI, decision structuring and making process, and organizational learning. Section 6 contains a discussion about the integration perspective of the decision-making process, according the scheme presented in Sect. 5. Finally, the conclusion presents the limitations of this study and highlights the insights this work has gained.

2 Big Data

With data increasing globally, the term "Big Data" is mainly used to describe large datasets. Compared with other traditional databases, Big Data includes a large amount of unstructured data that must be analyzed in real time. Big Data also brings new opportunities for the discovery of new values that are temporarily hidden [3].

Big Data is a broad and abstract concept that is receiving great recognition and is being highlighted both in academics and business. It is a tool to support the decision-making, process by using technology to rapidly analyze large amounts of data of different types (e.g., structured data from relational databases and unstructured data such as images, videos, emails, transaction data, and social media interactions) from a variety of sources to produce a stream of actionable knowledge [4].

After the data is collected and stored, the biggest challenge is not just about managing it but also the analysis and extraction of information with significant value for the organization. Big Data works in the presence of unstructured data and techniques of data analysis that are structured to solve the problem [1].

A combination called the 4Vs characterizes Big Data in the literature: volume, velocity, variety, and value [5]. Volume has a great influence when describing Big Data as large amounts of data are generated by individuals, groups, and organizations. Zikopoulus et al. reports that the estimated data production by 2010 was about 35 zettabytes [6].

The second item, velocity, refers to the rates at which Big Data are collected, processed, and prepared—a huge, steady stream of data that is impossible to process with traditional solutions, For this reason, it is important to consider not only "where" data are stored but also "how" they are stored.

The third item, variety, is related to the types of data generated from social sources, including mobile and traditional data. With the explosion of social networks, smart devices, and sensors, data have become complex because they include semi-structured and unstructured information from log files, web pages, index searches, cross-media, e-mail, documents, and forums.

Finally, the value can be discovered from the analysis of the hidden data, so Big Data can provide new findings of new values and opportunities to assist in making decisions. However, management of this data can be considered as a challenge for organizations [1].

In order to demonstrate the differentiation between Big Data and Small Data, we analyzed them using five main characteristics: goals, data location, data structure, data preparation, and analysis, in Table 1.

Importantly, relational databases are not obsolete, on the contrary, they continue to be useful to a number of applications. In practice, how larger a database becomes, the higher the cost of processing and labor, so it is necessary to optimize and add new solutions to improve storage providing greater flexibility.

For the purpose to better understand the impact of science and Big Data solutions, the applications and Big Data solutions in the following different contexts will be presented: education, social media and social networking, and smart cities.

Grillenberger and Fau used educational data to analyze student performance [7]. Their learning styles were also clarified by the use of Big Data in conjunction with teaching strategies to gain a better understanding of the students' knowledge and an assessment of their progress. These data can also help identify groups of students with similar learning styles or their difficulties, thus defining a new form of personalized learning resources based on and supported by computational models.

Big Data has created new opportunities for researchers to achieve high relevance when working in social networks. In this context, Chang, Kauffman and Kwon used communications environments to discuss the causes of the paradigm shift and explored the ways that decision support is researched, and, more broadly, applied to the social sciences [8].

In the context of a smart city, Dobre and Xhafa provide a platform for process automation collection and aggregation of large-scale information. Moreover, they present an application for an intelligent transportation system [9]. The application is designed to assist users and cities to resolving the traffic problems in big cities. The combination of these services provides support for the application in intelligent cities that can, benefit from using the information dataset.

The value of Big Data is driving the creation of new tools and systems to facilitate intelligence in consumer behavior, economic forecasting, and capital markets. Market domination may be driven by which companies absorb and use the best data the fastest. Understanding the social context of individuals' and organizations' actions means a company can track not only what their customers do but also get much closer to learning why they do what they do.

Table 1. Comparison of main characteristics of Big Data and Small Data.

Aspects	Big Data	Small Data
Goals	In general, they are projected from a predetermined goal and have a greater level of flexibility, considering the context of the problem. For example, the market scenario analysis to identify forms to accelerate the sales can be considered	They are generally designed to answer a specific question and control in a particular context. For example, inventory control, get only information concerning the entry and exit of goods, is not always done interaction between customer and supplier, acting on the basis of current market
Data location	The location normally aggregates data spread across different media, which can be in several Internet servers. The architecture consists of a distributed computing where multiple servers work together to store and process information. High power scalability, low cost of implementation	In general, the data come from the internal organization and the data files. For example, working with spreadsheets results in great increases on internal control
Data structure	The structure is usually able to absorb unstructured data (e.g., free text documents, images, movies, sound recordings, and physical objects). In others words, Big Data is just to be able to work with many variables simultaneously, as reading and rendering images in minimal time and very efficiently. For example, smart city applications, using real-time information to describe the traffic of a big city	The structure usually contains structured data. Data are represented by uniform records in an orderly spreadsheet. For example, the enterprise resource planning (ERP) that are systems that have a pre-defined architecture and their records represents a structured way to work with data within organizations
Data preparation	In general, the data come from different sources and are prepared by several users. People who use the data rarely are the ones who prepared. In this context, different people in different organizational roles contributes to disseminate information	In many cases, the data users prepare their own data for their own purposes. For example, presenting the results according to a specific context to which the user is located
Analysis	Analysis is usually done in incremental steps. The data are extracted, revised, normalized, processed, visualized, interpreted, and then analyzed with different methods. For example, complex techniques of data analysis combining data mining and artificial intelligence	In most cases, all the data for the project can be analyzed all at once. In this case, the structure is pre-defined and based on the specific context. Also are used Structured Query Language (SQL) combined with appropriate programming languages to create procedures to mining, process and analyze the data

Sources: [4, 1].

To date, for the use of Big Data, a modern infrastructure is needed to overcome the limitations related to language and methodology. Guidelines are needed in a short time in order to deal with such complexities, as different tools and techniques and specific solutions have to be defined and implemented. Furthermore, different channels through which data are collected daily increases the difficulties of companies in identifying which is the right solution to get relevant results from the data path. In this context, the technology of BI and DSS will be presented.

3   Decision Support Systems (DSS)

Information and knowledge are the most valuable assets for organizations' decision-making processes and need a medium to process data into information loaded with value and relevance for use in organizational processes. Information Systems (IS) represent these media. Specifically focused on the decision-making process, the DSS work for the processing, analyzing, sharing, visualizing of important information to aid in the process of knowledge aggregation and transformation, and thereby improve the organizational knowledge.

DSS are IS designed to support solutions for decision-making problems. The term DSS has its origin in two streams: the original studies of Simon's research team in the late 1950s and the early 1960s and the technical works on interactive computer systems by Gerrity's research team in the 1960s [10]. In a more detailed definition, DSS are interactive, computer-based IS that help decision-makers utilize data, models, solvers, visualizations, and the user interface to solve semi-structured or unstructured problems. DSS are built using a DSS Generator (DSSG) as an assembling component [11].

DSS have a strict link with intelligence-design-choice model, but acting with more power in the choice phase [2]. Their main objective is to support a decision by determining which alternatives to solve the problem are more appropriate. Although the choice is made by a human agent (a manager, treated as a decision-maker within this process), the DSS role is to provide a friendly interface where the agents can build scenarios and simulate and obtain reports and visualizations to support the decisions [12].

This kind of system has a set of basic elements that includes a data base and a model base with their respective management, the business rules to process data according a chosen model (e.g., the core of the system), and a user interface [10]. Data and model bases and their respective management systems allow for business rules in processing data according to a model to formulate the possibilities of solutions for the problem.

4   Business Intelligence

An organization's decision-making process begins with the intelligence phase of Simon's [2] model. In this phase, the perception is made that there is a problem to be solved in the future by applying problem structuring methods. Also in this phase, the tools of BI may be used to support the organization's discovery of opportunities for decision-making by, providing advanced analytics and assuring data integration [13].

So, besides the problem solving, decision opportunities can be added to the set of benefits that BI can bequeath to the scope of decision support.

The definition of BI can be expressed in two ways: in a holistic organizational decision-making approach and in the technical point of view [14]. Similarly Handzic, Ozlen and Durmic presents two kinds of concepts for BI: one centered on data analytics supporting decision-making processes in organizations and the other focused on tools and technologies for data storage and mining for knowledge discovery [15]. This study will adopt the strategy of considering BI as a system that includes both organizational and technical perspectives that provide the information needed to perform an analysis the aid the generation of decision opportunities, the decision-making process, and the organizational learning process.

Regarding this definition, BI covers all the processes involved in extracting valuable and useful information from the mass of data that exists within a typical organization to support the decision-making process. Business Intelligence Systems (BIS) are those that makes use of a combined process involving IT solutions and business experts' knowledge on the operation of business, integration and organizational management obtained as a result of intelligent decision-making [16, 17].

According to Azma and Mostafapour, there are two main features of data: the organizational learning process and the smart processing of data [17]. The organizational learning includes the discovery of new knowledge and dissemination of this knowledge to those who need it. The smart processing includes analyzing and assessing the information, providing decision support to ensure the aligning of the future performance of the organization with the planning, and providing knowledge feedback about the involved processes to be combined with pre-existing (explicit) knowledge.

Chang, Hsu and Shiau set up BI like both a product and a process. From the process perspective, the main goal of BI is to aid the decision-making process and reduce the time spent on the decision [18]. For this to occur, it is necessary that all the sets of basic components be defined and implemented. From the product perspective, the BI is the IT component that contains the referred to set of basic components and that can be used as the core engine of DSS to generate analytics for managers as the decision-makers.

From an organizational perspective, BI is part of a decision environment that combines both technology sets and human capacities in order to obtain decisions strategically aligned with the organization's planning. This is the holistic organizational decision-making approach that Işık, Jones and Sidorova mentioned and that still includes BI capabilities such as data quality, integration with other systems, user access, system flexibility, and risk management support [14]. These authors conceptualized the first three capabilities as technological BI and the last two as organizational BI.

Handzic reinforces the Knowledge Management (KM) and organizational learning point of view for BI, considering some organizational aspects focused on human interaction which the authors called socio-technical perspective: organizational culture, leadership, and measurement for successful implementations [19]. Figure 1 shows this perspective.

Considering the definitions by Handzic [19] and Grünwald and Taubner [20] leads to understanding that the information evolutionary scale is made possible through a technical toolset that supports the necessary transformations between the points of the

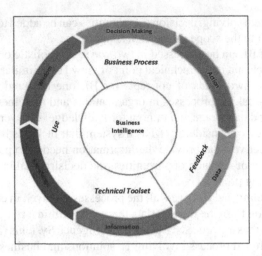

Fig. 1. Business process support using BI. Source: Adapted from [20].

scale (since data is until wisdom), including elements of Artificial Intelligence (AI). The business process, in turn, includes the decision-making process and the action to implement the decision. In Fig. 1, the combination of BI, business process, and the technical toolset could be understood as the elements that define a DSS, because they represent information transformations followed by their application in decision-making and the provision of feedback to initiate a new cycle of information transformation.

The works cited previously describe intrinsic issues to Big Data and highlight the use of BI alone. The differential of our proposal is to get an integrated view of the technologies involved, aggregating value to the decision-making process.

5   Integrated Model for Decision-Making Process, Big Data, and BI Tools

Simon's decision model summarizes the decision-making process into three phases, as introduced previously. Each phase this model is susceptible to the use of methods and tools from organizational and technological perspectives. The organizational perspective may use Problem Structuring Methods (PSM); Multi-criteria Decision Aid (MCDA); and KM techniques such as brainstorming, communities of practice, best practices, narratives, yellow pages, peers assistance, and knowledge mapping. These methods and techniques aids in the knowledge elicitation of the actors involved in the decision-making process, thus contributing to identify the necessary expertise necessary for solve the specific problem in question, as in the case of PSM and KM techniques, or acting to provide recommendations to solve this problem, as in the case of MCDA.

Technological tools involve data repositories (e.g., data warehouses and data marts) filled with data from public sources, BI or even AI and Problem Solving Methods (PSolM) originated from Knowledge Engineering (KE) (e.g., CommonKADs and Methodology and Tools Oriented to Knowledge-Based Engineering Applications [MOKA]).

These tools are important elements that contribute to store, access and analyze information, discovering and sharing knew knowledge in databases and even supporting the application of the organizational perspective's methods and technics.

The main purpose of this work is the integration of the decision-making process with some of these tools presented in Fig. 2, considering the perspective of the predictive approach to decision-making. In this perspective, the use of methods to structure decision problems and suggest alternatives to choose from is an important issue and an efficient way to support the DSS design and development. Combined with the predictive approach, this process makes use of BI tools to provide domain information to aid all the phases of the process.

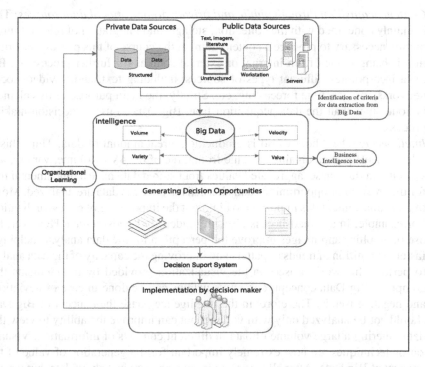

Fig. 2. Integrated model of the decision-making process.

It is noteworthy that some of these elements are framed within the phases of Simon's model. In the phase of intelligence, by making use of Big Data powered by internal and external data sources, organizations can make use of BI strategies and tools to aid in identifying relevant information, and then the generation of decision opportunities occurs.

The function of the design phase is to provide a methodology to aid the choice of the alternatives based in what was defined in the problem structuring process during the intelligence phase. This design must also be incorporated into this methodology, as formal aspects related to the method or model that are defined according the problems identified during the problem structuring process. The development of DSS has made the use of this model viable by allowing the decision-makers, through a friendly and easy-to-use interface, to perform a series of configurations.

In the final phase of choice, the decision-makers will use the results generated by DSS to complete the decision-making process with the choice of one, or a set of, alternatives, that will then be implemented by the organization.

All these processes produce new knowledge to be combined with previous knowledge about the domain of the problem. This new knowledge will provide feedback to power the Big Data so that it can be used as necessary, thus fulfilling its role in the organizational learning process.

Each element of the integrated model is described as follows:

(a) *Content acquisition through public and private organizational data sources:* This is mainly concerned with the collection, storage, and integration of relevant information necessary to produce a content item. In the course of this process, information is being pooled from internal or external sources for further processing. Big Data incorporates different types of sources, including text, audio, video, social networks, images, time forecasting, etc. Strictly, the main purpose of this element is none other than the data acquisition from Big Data to use in decision-making process.

(b) *Intelligence:* The whole world is producing a great amount of data. Thus, this is relevant as Big Data obtains its value from three of the 4Vs: volume, variety, and velocity. In this phase, aggregated values from stored data have a fundamental role for the creation of opportunities and alternatives once the data are analyzed. Moreover, in this context it is important to highlight the importance of data visualization. For example, in a spreadsheet is difficult to identify trends in data. However, the use of graphics and images improve the perception for the data analysis helping a faster recognition of trends or patters and improving the capacity of the data analyst to perform his work. Based on the visualization provided by the elements that composes Big Data concept, corrective actions can be done in case of deviations and negative trends. Therefore, in the intelligence phase the concept of Big Data should not be analyzed only with volume, but can improve the ability to view this data, filtering a large volume of data in different contexts of information. Visualization techniques are now extremely important for the generation of value of the concept of Big Data. After all, Big Data is not a concept just about data, but we can extract insights and intelligence and visualization is the fundamental key to the decision-making process. The intelligence represents the capacity to aggregate value to acquired data in order to obtain relevant information, applicable in the organizational problem solving. These information should be capable of contextualize with internal and external phenomena of the organization, ensuring the other following elements the necessary power of action to satisfactorily contribute to resolve the problem.

(c) *Opportunities and alternatives generation:* This is the process of creating alternatives, which is not a trivial task. It starts with dataset analysis that enable decision-makers to obtain a global view of the process. Then, from the analyses performed through BI tools with Big Data content, decision-makers pro-actively create opportunities and generate opportunities to solve the decision problem. This phase also works for the definition of the criteria, which the decision-makers will use to judge or evaluate each alternative.

(d) *DSS:* With the opportunities identified and having the criteria and alternatives to evaluate, DSS may be implemented according a decision problem that will predict which method is the most adequate. DSS will act in helping decision-makers in obtaining an indication or a recommendation of alternatives to choose from that will be implemented to solve the problem.

(e) *Implementation of decision:* After a choice is made, alternatives will be implemented in organizations to actively solve the identified problem.

As a last element, the Organizational Learning says respect to all these processes' elements generating important knowledge about the decision problem. This knowledge may be captured, registered, and stored in a knowledge repository to provide organizational memory about the problem domain and will be available for use at any time. The standard flow of this new knowledge, after the implementation of the chosen action, runs to private (or internal) data sources, e.g., a base of managerial practices.

6 Discussion

Knowledge extracted adequately from Big Data aggregates the value that decision-makers use to identify a decision opportunity. This work provided theoretical evidence to corroborate the idea that the perspective of historical data combined with decision-makers' knowledge and experience, formal problem structuring, and use of decision methods or models may make the decision-making process more robust and more reliable.

Generally, companies use the descriptive approach to make decisions, by performing an analysis based only on historical data. The focus solely on the past makes it difficult to concentrate on new strategies for the future. The proposition of the present work also considers this descriptive approach, but it recognizes the value of the predictive approach in order to provide recommendations to solve a decision problem, based on decision-makers' knowledge and judgment, and information technology: Big Data, BI, and DSS.

The Big Data study performed here started with the analysis of the data's influence over the decision-making process by ensuring that decision-makers can discover opportunities to act problem solving.

The main contributions of the theoretical approach presented here are (i) develop a perspective that combines the decision-making process, Big Data, BI, DSS, and organizational learning and (ii) use the concept that Big Data works as a data provider over which may be applied BI techniques and tools may be applied mainly in supporting the discovery of opportunities for a decision.

Decision-makers, when preparing for making a decision, incorporate their knowledge and discernment along with an organizational learning process that will help them to create an organizational memory that provides knowledge generated through the process for later use. Thus, beyond technological toolsets and decision-making and methodologies, the process described here takes into account the subjective characteristics linked to the decision-makers' perceptions, experiences, and personalities.

The use of Big Data provides to managers the possibility to explore both internal and external information, not only identifying a decision problem but also having as proposal the potential to increase de intelligence power within the decision-making process.

7 Conclusion

The increasing amount of data that arrives at organizations accumulate through electronic communication is amazing, in that not only has the volume of the data change, but also the variety of information collected in through several communication channels ranging from clicks on the Internet to the unstructured information from social media. In addition, the speed at which organizations can collect, analyze, and respond to information in different dimensions is increasing.

Big Data has become a generic term, but in essence, it presents two challenges for organizations. First, business leaders must implement new technologies and then prepare for a potential revolution in the collection and measurement of information. Second, and most important, the organization as a whole must adapt to this new philosophy about how decisions are made by understanding the real value of Big Data.

Organizations must understand the role of the Big Data associated with decision-making, with the emphasis on creating opportunities from these decisions, because we live in a world that is always connected, and where consumer preferences change every hour. Thus, analysts can check multiple communication channels simultaneously and trace certain profiles or decider behaviors.

The main contribution of this work is to promote the integrated view of Big Data, BI and DSS inside the context of decision-making process, assisting managers to create new opportunities to resolve a specific problem.

The crucial point is to look widely for new sources of data to help make a decision. Furthermore, Big Data not only transforms the processes of management and technology but it also promotes changes in culture and learning in organizations.

Ultimately, Big Data can be very useful if used adequately in the decision-making process, but just its use will not guide the decision itself and it will not generate alternatives or predict the results. For this, the participation of decision-makers is essential, as their experience and tacit knowledge are necessary to aggregate value over information and the possible knowledge stored.

From this initial study, where the idea of get an integrated view of all these elements as decision-making tools, we can create a set of perspectives to apply in future researches, as example a detailed exploration focused on each phase of the model. Other ideas: semantic exploration of Big Data applied to decision problems structuring, direct integration between Big Data and BI tools to fulfil organizational repositories providing data to the information systems.

References

1. Chen, M., Mao, S., Liu, Y.: Big data: a survey. Mob. Netw. Appl. **19**, 171–209 (2014)
2. Simon, H.A.: The New Science of Management Decision. Harper and Row, New York (1960)

3. Renu, R.S., Mocko, G., Koneru, A.: Use of big data and knowledge discovery to create data backbones for decision support systems. Procedia Comput. Sci. **20**, 446–453 (2013)
4. Berman, J.J.: Principles of Big Data Prepararing, Sharing, and Analyzing Complex Information. Elsevier, Waltham (2013)
5. Klein, D., Tran-Gia, P., Hartmann, M.: Big data. Informatik-Spektrum **36**, 319–323 (2013)
6. Zikopoulus, P.C., Eaton, C., deRoos, D., Deutsch, T., Lapis, G.: Understanding Big Data: Analytics for Enterprise Class Hadoop and Streaming Data. McGrow Hill, New York (2012)
7. Grillenberger, A., Fau, F.E.: Big data and data management: a topic for secondary computing education. In: ICER 2014 – Proceedings of the 10th Annual International Conference on International Computing Education Research, pp. 147–148 (2014)
8. Chang, R.M., Kauffman, R.J., Kwon, Y.: Understanding the paradigm shift to computational social science in the presence of big data. Decis. Support Syst. **63**, 67–80 (2014)
9. Dobre, C., Xhafa, F.: Intelligent services for big data science. Futur. Gener. Comput. Syst. **37**, 267–281 (2014)
10. Liu, S., Duffy, A.H.B., Whitfield, R.I., Boyle, I.M.: Integration of decision support systems to improve decision support performance. Knowl. Inf. Syst. **22**, 261–286 (2009)
11. Dong, C.-S.J., Srinivasan, A.: Agent-enabled service-oriented decision support systems. Decis. Support Syst. **55**, 364–373 (2013)
12. Daas, D., Hurkmans, T., Overbeek, S., Bouwman, H.: Developing a decision support system for business model design. Electron. Mark. **23**, 251–265 (2012)
13. Popovič, A., Hackney, R., Coelho, P.S., Jaklič, J.: Towards business intelligence systems success: effects of maturity and culture on analytical decision making. Decis. Support Syst. **54**, 729–739 (2012)
14. Işık, Ö., Jones, M.C., Sidorova, A.: Business intelligence success: the roles of BI capabilities and decision environments. Inf. Manag. **50**, 13–23 (2013)
15. Handzic, M., Ozlen, K., Durmic, N.: Improving customer relationship management through business intelligence. J. Inf. Knowl. Manag. **13**, 1450015-1–1450015-9 (2014)
16. Mohamadina, A.A., Ghazali, M.R.B., Ibrahim, M.R.B., Harbawi, M.A.: Business intelligence: concepts, issues and current systems. In: 2012 International Conference on Advanced Computer Science Applications and Technologies, pp. 234–237 (2012)
17. Azma, F., Mostafapour, M.A.: Business intelligence as a key strategy for development organizations. Procedia Technol. **1**, 102–106 (2012)
18. Chang, Y.-W., Hsu, P.-Y., Shiau, W.-L.: An empirical study of managers' usage intention in BI. Cogn. Technol. Work **16**, 247–258 (2013)
19. Handzic, M.: Integrated socio-technical knowledge management model: an empirical evaluation. J. Knowl. Manag. **15**, 198–211 (2011)
20. Grünwald, M., Taubner, D.: Business intelligence. Informatik-Spektrum **32**, 398–403 (2009)

Cloud Enabled Big Data Business Platform for Logistics Services: A Research and Development Agenda

Irina Neaga[1](✉), Shaofeng Liu[1], Lai Xu[2], Huilan Chen[1], and Yuqiuge Hao[3]

[1] Graduate School of Management, Plymouth University, Plymouth, UK
{irina.neaga,shaofeng.liu,huilan.chen}@plymouth.ac.uk
[2] Bournemouth University, Bournemouth, UK
lxu@bournemouth.ac.uk
[3] University of Vaasa, Vaasa, Finland
yuqiuge.hao@uwasa.fi

Abstract. This paper explores the support provided by big data systems developed in the cloud for empowering modern logistics services through fostering synergies among 3/4PL (third /fourth party logistics) in order to establish interoperable or highly integrated and sustainable logistics supply chain services. However, big data applications could have limited capabilities of providing performant logistics services without addressing the quality and accuracy of data. The main outcome of the paper is the definition of an architectural framework and associated research and development agenda for the application of cloud computing for the development and deployment of a Big Data Logistics Business Platform (BDLBP) for supply chain network management services. The capabilities embedded in the BDLBP can provide powerful decision support to logistics networking and stakeholders. Two of the three strategic and operational capabilities as operational capacity planning, and real-time route optimisation are built upon literature based on operational research, and are extended to the scope of dynamic and uncertain situations. The third capability, strategic logistics network planning is currently under researched and this approach aims at covering this capability supported by big data analytics in the cloud.

Keywords: Big data · Cloud computing · 3/4PL · Big data logistics business platform (BDLBP) · Business intelligence · Big data analytics

1 Introduction

Logistics and supply chain network management have evolved considerably in last few decades. The most significant advancements in modern logistics are undoubtedly the emergence of third party logistics (3PL) and more recently the fourth party logistics (4PL). The approach proposed in this paper aims to investigate new means to empowering the capabilities of modern logistics network formed by 4PL based on big data systems in the cloud. Therefore all types of data (such as customer data, commercial data, operations data, marketing and sales data, product data, logistics network data, and real-time transportation traffic and incident data) can be coherently integrated and

© Springer International Publishing Switzerland 2015
B. Delibašić et al. (Eds.): ICDSST 2015, LNBIP 216, pp. 22–33, 2015.
DOI: 10.1007/978-3-319-18533-0_3

analysed to better support 4PL's strategic and operational logistics decision making. This research is mainly directed to the development of an innovative **Big Data Logistics Business Platform (BDLBP)** that can provide new solutions to support improved performance with new capabilities such as 4PL's strategic logistics network design, operational capacity planning, route planning and optimisation, risk evaluation and resilience planning, customer personalisation and loyalty management, and emerging businesses (e.g. real-time delivery route optimisation and crowd-based pickup and delivery). The innovation of the BDLBP is that it can provide seamless integration of all types of logistics data that may come from various sources, typically **human-sourced data** including data from social networks, **process mediated data** such as data embedded in logistics business processes and transactions, and **machine-generated data** such as data from sensors and machines such as smart phones and GPSs employed to measure and record the events and situations in physical logistics world. More importantly, the BDLBP will enable the conversion of human-sourced data and machine-generated data into process-mediated data in real-time or near real-time to improve 4PL's logistics business decision making, in order to transform the data power into the logistics business value. It will be also possible to apply big data analytics based on information mining and knowledge discovery for better decision making and forecasting based on large sets of historical data. However the lack of data accuracy and quality might hinder the contribution of big data analytics to logistics performance. Also decision making systems should be designed using operational research methods and big data exploration will enhance the performance of logistics business decision making. Many companies participating in logistics activities have applied big data to support their business. However, most of the applications remain within the classic in-house logistics and the scope of the logistics activities is very limited. According to the Supply Chain Big Data Report compiled from the transportation perspective [1], over 90 % of the companies which are implementing big data supply chain analytics are retailers and manufacturers undertaking logistics (retailers 44.7 %, consumer goods manufacturers 22.3 %, high-tech manufacturers 12.6 % and pharmacy manufacturers 10.7 %). Studying and applying big data for outsourcing logistics providers such as 4PL and 3PL is currently under researched. One possible explanation for this could be that employing big data analytics in 4PL and 3PL could be a lot more difficult than in-house logistics conducted by retailers and manufacturers, because data management and communication for 4PL and 3PL need to involve a lot more stakeholders, especially the extension of supply chain from vertical integration to horizontal integration.

1.1 Conceptual Approach

A 3PL service provider is an external supplier that performs part or all of the logistics functions of companies, including transportation, warehousing, distribution, collection, packaging, documentation, and so on [2, 3]. However, with the development of outsourcing and the rising expectations on product variety and quality and service level of customers, unique and comprehensive supply chain solutions can no longer be achieved by a single 3PL provider. Thus, a new generation of logistics provider, 4PL, has emerged as a breakthrough solution to meet modern logistics challenges [4]. The initial concept of

4PL was introduced by Accenture, which functions as an integrator that assembles the resources, capability, and technology of its own organisation and other organisations to design, build, and run comprehensive, concerted supply chain solutions [5, 6]. It has been gradually proven that the 4PL concept can excel 3PL as the logistics new business model and will become the dominant development direction of the logistics industry either in theory or in practice for the next decade [6, 7].

Over the past years, many researchers have been endeavouring in leveraging logistics performance with the advancement and power from new ICT technologies, both in general [8, 9] and most notably in semantics and ontology for logistics [10–14] agent technology [15, 16], M2M and Internet of things [17, 18], RFID, wireless networks and wireless sensor networks [19–22], Cloud based and Self-Organizing networks and structures [23–25], and business intelligence in logistics supported by soft computing techniques [26–32]. During this paradigm shift from the management of static to mobile, and from small amount of individually to vast amount of socially created information, big data is considered to have great potential to transform the logistics industry. It is widely accepted that big data are emerged and that could provide a real support for logistics services [32] for supply chain network management. Generally big data is characterized by 3Vs (dimensions) [33]: (1) volume: processing large amounts of data, more than terabytes or petabytes; (2) velocity: processing in real- or near real-time, in batches or as streams; (3) variety: data coming from a great variety of heterogeneous sources (structured, unstructured or semi-structured). In addition, another 2Vs have been recently identified: (4) veracity (of data sources): data consistency (or certainty) which is strongly related to statistical reliability, and data trustworthiness which depends on data origin, collection and processing methods, including trusted infrastructure and facility; and (5) value (economic and social outcomes): added-value that the collected data can bring to the intended process or activity. This model is shown in Fig. 1.

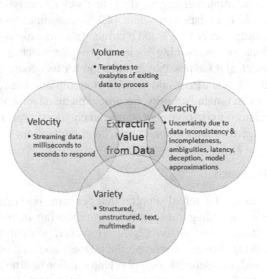

Fig. 1. Big data model (Revised of the source: [33])

It is recognised that big data could support modern logistics represented by today's 4PL and 3PL [33]. Even though there has been plenty of discussion on conceptual approach of using big data for logistics industry, there has been so far no mature systematic frameworks and software platforms based on big data that can support 4PL to achieve its ambition of integrating resources, capabilities and technologies across supply chains to deliver concerted, comprehensive, location-based services to customers at the convenience of their doors at the times they wish.

Today's 4PL providers have to effectively manage a massive flow of goods to achieve the best performance and at the same time create vast data sets. Therefore, logistics managers rely on the business intelligence with the aim of transforming raw data into meaningful and useful information for decision-making. Business intelligence comprises computing technologies for the identification, discovery, and analysis of data. Handling properly the huge amount of available data poses several challenges that need to be addressed by the big data community. For example, 4PL may need to integrate the right resources, capabilities and technologies to deliver thousands of shipments every day, of which the origin and destination, size, weight, content and location etc. need to be tracked across global delivery networks. Modern logistics such as 4PL therefore faces a number of key challenges: gathering data from information silos within different stakeholders – inability to "connect the dots" between data sources; lack of logistics domain knowledge around the data – inability to analyse the data in line with logistics business processes; lack of an integrated management approach for targeted audience via various views – inability to personalise and customise logistics services.

1.2 Cloud Computing Supporting Big Data Applications

The US National Institute of Standards and Technology (NIST) provides a definition of cloud computing as "… a model for enabling ubiquitous, convenient, on-demand network access to a shared pool of configurable computing resources (e.g., networks, servers, storage, applications, and services) that can be rapidly provisioned and released with minimal management effort or service provider interaction…" [34] Cloud leverages virtualization technology to achieve the aim of providing computing resources as a utility [35]. It dramatically reduces the complexity of owning infrastructure, software and services through on-demand resources provisioning over virtualization and pay-as-you-go usage pattern. In cloud computing, data and software applications are defined, developed and implemented as services. Different services provide different types of functionality to different consumers. These services have been defined as a multi-layered infrastructure (as shown in Fig. 2) and are described as follows [36, 37]:

1. *Software as a Service* (SaaS), where applications are hosted and delivered online via a web browser offering traditional desktop functionality
2. *Platform as a Service* (PaaS), where the cloud provides the software platform for systems (as opposed to just software)
3. *Infrastructure as a Service* (IaaS), where a set of virtual computing resources, such as storage and computing capacity, are hosted in the cloud; customers deploy and run only their own applications for obtaining the needed services.

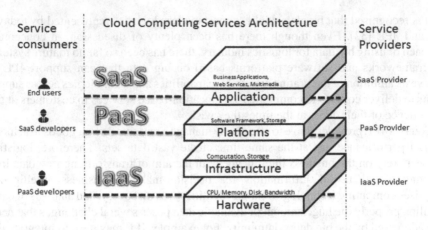

Fig. 2. Cloud computing model and services architecture (Revised based on [35])

Figure 2 illustrates the service layers, services providers and consumers. It shows the entire business model of cloud computing that could be used for big data solutions. The hardware and platform-level resources are provided as services on an on-demand basis. Each layer can be implemented as a service to the layer above [35]. For instance, PaaS runs its service on top of IaaS's service. However, in practice, the PaaS provider and IaaS provider are often parts of the same organization that is defined as the infrastructure provider. Big data strategies using cloud services are rapidly advancing [38]. Cloud computing is considered the key technology for big data applications and analytics as presented in Fig. 3; not only because it provides an infrastructure and systems, but also because it is a business model that big data analytics can adopt (e.g. Analytics as a Service (AaaS) or Big Data as a Service (BDaaS) as well as knowledge and information as services (KaaS), (IaaS)).

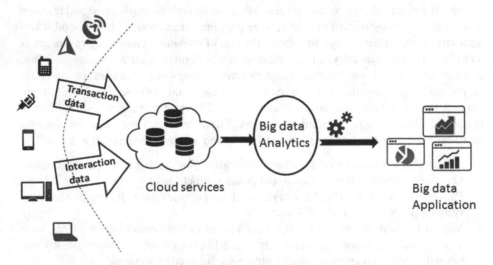

Fig. 3. Big Data analytics and applications in the cloud

2 Big Data Logistics Business Platform (BDLBP)

The paper proposes a platform that applies the service paradigm associated with cloud design principles having the foundation in service architectures. Big data, information and knowledge are created using services that support flexible logistics processes.

In the creation of these services, the platform has the capability to flexibly integrate both structured and unstructured data from multiple sources stored in the cloud. The design and creation of the services is enabled by understanding of the knowledge gained through analysis, visualisation and data manipulation. The resulting services are easily invoked by logistics business users to form higher level supply chain services or e-commerce services that may be used in mobile and web environments.

Specific principles are used to design the BDLBP architectural framework such as:

- Logistics information links aim to bridge the gap between various sources and types of logistics information and users of different backgrounds (e.g., even non-IT savvy user) by providing an integrated information source.
- The integration and interoperability support different data proprietary, legacy and existing solutions through open standards for logistics systems.

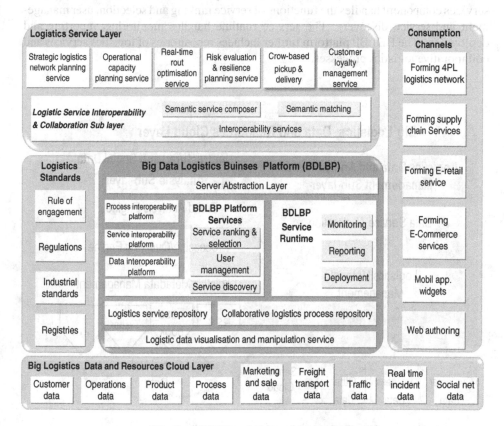

Fig. 4. BDLBP architectural framework

- A standardised data model aims to develop a unify description language for logistics data and information to enable the integration and delivery of logistics data in a unified manner.
- Service infrastructure aims to provide an infrastructure towards integration and provision of logistics related information processing as services. The infrastructure applies service oriented architecture (SOA) and cloud principles.
- Service consumption based on the advancement of Web 2.0 technologies, the delivery channel of logistics services should incorporate the collaborative, bottom-up principles of service consumption.
- Organisational impact requires a cross-organisational, collaborative attitude to the design, prototyping, deployment and test.

Figure 4 presents the architectural framework. The overall architecture is organised into five parts: Big Data Logistics Business Platform, Logistics Service Layer, Logistics Standards, Big Logistics Data and Resources Cloud Layer, and Consumption Channels.

At the core of this architectural framework the **big data logistics business platform (BDLBP)** is defined as a core integration approach. It delivers the necessary extensions and developments of the logistics data visualisation and manipulation services, supports data, service and logistics process interoperability. Within the BDLBP, the platform services component handles the functions of service ranking and selection, user management and service discovery. The service runtime handles monitoring, reporting and deployment. The BDLBP platform further includes repositories of logistics services and collaborative logistics processes.

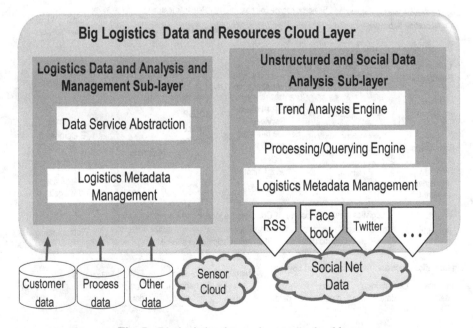

Fig. 5. Big logistics data and resource cloud layer

These repositories enable sharing and reuse of existing logistics solutions. The **logistics service layer** contains various logistics capability services and a sub-layer for logistics service interoperability and collaboration. The logistics service interoperability and collaboration sub-layer enables a semantic service composer, semantic matching and interoperability services. The **logistics standards** allows transforming textual rule of engagement, regulations of different countries, and industrial standards into computer understandable knowledge. This can then be used to ensure that new forming logistics processes or services comply with existing rules. The **consumption channels** depict the various ways in which the architecture can be used by external parties. Envisioned uses include: forming 4PL logistics networks; forming supply chain services; forming e-retail services; forming e-commerce services; mobile applications and web components; and web authoring.

The **big logistics data and resources cloud layer** handles data from different resources. Figure 5 shows detailed components of the big logistics data and resources cloud layer.

3   Development Challenges

In order to effectively develop and apply the BDLBP additional in depth research on big data strategies for modern 3/4PL logistics services is needed. This approach suggests developing new technology based on existing gaps in technology and provide directions on further research on cloud enabled big data solutions for modern logistics. The cloud services for big data integration will be addressing not just the big data identification and /or acquisition, but also the integration of dissimilar logistics data from web, social networks and legacy logistics systems. The logistics data, whether human-sourced or machine-generated, will be transformed into process-mediated data and be integrated into the 4PL business processes. Hence research underpinning big data strategies with logistics business process intelligence, i.e. data with production power to create value to logistics business. The use of semantics and dedicated logistics ontology will support the heterogeneous data acquisition, integration and presentation.

The capabilities designed inside the BDLBP can provide powerful decision support to logistics stakeholders. Two of the three strategic and operational capabilities (operational capacity planning, and real-time route optimisation) are built upon literature, but are extended to the scope of dynamic and uncertain situations. Existing work on operational capacity planning and route optimisation is restricted to quantitative analysis based on static data [6]. Further research will explore the big data approaches considering the velocity of big data by applying logistics data collected in real-time, and hence advance existing work by promptly capturing and processing constantly changing logistics data for decision support. Further approach will investigate advanced data management capabilities that will provide accurate data of high quality. The third capability, strategic logistics network planning is currently under researched. Because of the lack of data communication standards across different regions, countries and continents, integration of data for example collected by embedded devices such as RFID and GPSs still remains unresolved. It is not uncommon that in international logistics, goods items and parcels get unrecognised after they enter into a region, country or continent where RFID devices use different data standards [39]. Logistics process and service interoperability is another key issue. Due to its human-centric character, interoperability plays a key role in the area of intelligent decision support systems where there is a huge interaction with humans, as in the context of logistics services. The behaviour exhibited by an interpretable system must be easily understandable, explicated or accounted for a human being. Subsequently, such systems endowed with interpretability capabilities are likely to be trusted by end-users, increasing the success rate of introducing intelligent systems into the market and contributing to customer loyalty management. Existing research has mainly addressed the interoperability issue at data level. Embedded within the platform BDLBP include three core functions related to interoperability. These are process interoperability, service interoperability and data interoperability. By achieving the process and service interoperability, BDLBP platform could provide logistics providers and customers a more concerted logistics business process and better integrated logistics services. This again highlights the strength of BDLBP in transforming data power into production (process) power and improving customer (service) experience. The application of big data mining and analytics using cloud services [38] will empower the platform with additional capabilities. The implementation solutions could be based on Google MapReduce, Hadoop Reduce, Twister, Hadoop ++, Haloop, Spark etc. which are used to process big data and run computational tasks as services. According to Assunção et al. (2014) [38] in the context of big data analytics, MapReduce represents an interesting model where data locality is explored to improve the performance of applications. Hadoop, an open source MapReduce implementation, allows for the creation of clusters that use the Hadoop Distributed File System (HDFS) to partition and replicate data sets to nodes where they are more likely to be consumed by mappers. In addition to exploiting concurrency of large numbers of nodes, HDFS minimises the impact of failures by replicating data sets to a configurable number of nodes. The cloud databases are used to store massive structured and semi-structured data generated from different types of applications. The most important cloud databases strategies include the BigTable, Hbase and HadoopDB. In order to implement an efficient big data mining and analysis framework the data warehouse processing is also important and the most important data warehouse processing technologies include the Pig, Hive etc. This research will take the technology beyond the state of the art in data and knowledge visualisation for 4PL users. Through the development of the new technologies for visual analytics, decision dashboards and portals, and data and knowledge visualisation, both 4PL stakeholders and end users can have personalised and customised user interfaces to the BDLBP, easily access the platform based on their background knowledge of logistics and their preferences in using logistics data. BDLBP becomes a communication platform for different users who do not have to be competent IT experts but can all search for the data, information and knowledge most relevant to their role in 4PL global network, and all collaborate closely facilitated by the coherent logistics business process embedded in the platform.

4   Concluding Remarks

Cloud enabled big data applications in modern logistics provide the opportunities for the redesign of the global logistics processes across. Logistics processes highly rely on the available logistics resources and technologies. When the barrier between different stakeholders such as authorities, shippers, all logistics providers including 3PLs and 4PLs, and others, is overcome by real-time and near real-time share of data and information, reconfiguration of logistics resources and technologies from different stakeholders to re-design logistics processes can then be possible in real-time or near real-time. This ongoing research supports the most state-of-the-art 4PL that aims to assemble the logistics resources at global level, i.e. to foster an open global network. All independent supply networks can then use the BDLBP as a common communication platform and be plugged into the platform as an agile service agent. The platform could hence enable the smooth flow of information and goods between the independent supply networks and the global network. Big data logistics business platform (BDLBP) will establish consistent data/information/knowledge exchange and communication strategies to foster synergies through:

a. *"improved collaboration and concerted actions between all authorities, shippers, retailers and logistics service providers for door-to-door transport"*. The BDLBP, through which all logistics stakeholders, including the authorities, shippers, retailers and logistics providers (4PL and 3PLs) can share logistics data and information across the 4PL global network, monitor and report current order information, customer information, traffic information and incident information etc. further to deploy delivery resources to better meet the delivery requirements. By using the BDLBP, the 4PL integrator hence can better co-ordinate the resources, technologies and capabilities plugged in the platform to improve collaboration and concerted actions between all stakeholders.
b. *"co-operative intelligent transport systems and cloud based services, integrated into one online planning platform that offers new means of communication among vehicles, between delivery vehicles and traffic management and to end users"*. The BDLBP platform integrates logistics data and cloud resources addressing the interoperability across logistics data, logistics processes and logistics services. All users of the BDLBP, including delivery vehicles, traffic management and end users, can effectively communicate with each other. Therefore, logistics strategic and operational planning can be performed through the BDLBP platform in real-time or near real-time.

References

1. Reynalds, S.: Supply chain executives weigh in: investment plans, solution sourcing and implementation challenges. Supply Chain Big Data Report. Eye for Transport (2013)
2. Sohail, M.S., Sohal, A.S.: The use of third party logistics services: a Malaysian perspective. Technovation **23**(5), 401–408 (2003)

3. Marasco, A.: Third-party logistics: a literature review. Int. J. Prod. Econ. **113**(1), 127–147 (2008)
4. Win, A.: The value a 4PL provider can contribute to an organisation. Int. J. Phy. Distrib. Logistics Manag. **38**(9), 674–684 (2008)
5. Chen, K.H., Su, C.T.: Activity assigning of fourth party logistics by particle swam optimisation-based pre-emptive fuzzy integer goal programming. Expert Syst. Appl. **37**, 3630–3637 (2010)
6. Yao, J.M.: Decision optimisation analysis on supply chain resource integration in fourth party logistics. J. Manufact. Syst. **29**(4), 121–129 (2010)
7. Warrilow, D., Beaumont, C.: 3PLs vs. 4PLs the great debate. Logistics Transp. Focus **9**(6), 30–33 (2007)
8. Adam, U., Tan, M.I.I., Desa, M.I.: Logistics and information technology: previous research and future research expansion. In: The 2nd International Conference on Computer and Automation Engineering (ICCAE), vol. 5, pp. 242–246 (2010)
9. Leung, S.C.H., Lim, M.K., Tan, A.W.K., Yu, Y.K.: Evaluating the use of IT by the third party logistics in South East Asia to achieve competitive advantage and its future trend. In: 8th International Conference on Information Science and Digital Content Technology (ICIDT), vol. 2, pp. 465–469 (2012)
10. Lian, P., Park, D.W., Kwon, H.C.: Design of logistics ontology for semantic representing of situation in logistics. In: Second Workshop on Digital Media and its Application in Museum & Heritages, pp. 432–437 (2007)
11. Liou, W.C., Chang, J.Y.: Multi-view ontology based logistical management system. J. Glob. Bus. Manag. **4**(1), 7–18 (2012)
12. Hoxha, J., Scheuerman, A., Bloehdorn, S.: An approach to formal and semantic representation of logistics services. In: Workshop on Artificial Intelligence and Logistics (AILog) at the 19th European Conference on Artificial Intelligence (ECAI 2010), Lisbon, Portugal (2010)
13. Anand, N., Yang, M., van Duin, J.H.R., Tavasszy, L.: GenCLOn: an ontology for city logistics. Expert Syst. Appl. **39**(15), 11944–11960 (2012)
14. Scheuermann, A., Hoxha, J.: Ontologies for intelligent provision of logistics services. In: The Seventh International Conference on Internet and Web Applications and Services (2012)
15. Gou, H., Uddin, M.K., Hossen, M.K.: An agent-based cooperative communication method in wireless sensor network for port logistics. In: 13th International Conference on Computer and Information Technology (ICCIT), pp. 494–499 (2010)
16. Tamagawa, D., Taniguchi, E., Yamada, T.: Evaluating city logistics measures using a multi-agent model. Procedia – Soc. Behav. Sci. **2**(3), 6002–6012 (2012)
17. Chen, D., Chang, G., Li, J., Jia, J.: Study on the interconnection architecture and access technology for Internet of Things. In: International Conference on Computer Science and Service System (CSSS), pp. 1744–1748 (2011)
18. Vandikas, K., Liebau, N.C., Dohring, M., Mokrushin, L. Fikouras, I.: M2M service enablement for the enterprise. In: 15th International Conference on Intelligence in Next Generation Networks (ICIN), pp. 169–174 (2011)
19. Li, B., Li, W.: Logistics information fusion application research based on RFID and GPS. In: The 27th Chinese Control Conference, pp. 389–393 (2008)
20. Marchese, M.: Wireless pervasive networks for safety operations and secure transportations. 5th IEEE International Symposium on Wireless Pervasive Computing, pp. 226–231 (2010)
21. Jang, L.G., Yang, S.F., Ho, T.S., Li-Yen Lai, L.Y., Nien, C.C: Logistics information monitoring by means of RFID sensor tag. In: International Conference on Information Management, Innovation Management and Industrial Engineering, vol. 3, pp. 86–89 (2012)

22. Zoller, S., Reinhardt, A., Steinmetz, R.: Distributed data filtering in logistics wireless sensor networks based on transmission relevance. In: IEEE 37th Conference on Local Computer Networks (LCN), pp. 256–259 (2012)
23. Viana, A.C., Mitton, N., Schmidt, L., Vecchio, M.: A k-Layer self-organizing structure for product management in stock-based networks. In: IEEE 7th International Conference on e-Business Engineering (ICEBE), pp. 198–205 (2010)
24. Gao, J., Ma, J., Zhang, X., Lu, D.: Cloud computing based logistics resource dynamic integration and collaboration. In: IEEE 16th International Conference on Computer Supported Cooperative Work in Design (CSCWD), pp. 939–943 (2012)
25. Arnold, U., Oberlander, J., Schwarzbach, B.: Advancements in cloud computing for logistics. In: Federated Conference on Computer Science and Information Systems, pp. 1055–1062 (2013)
26. Zimmermann, H.: Computational Intelligence in Logistics. In: Fogel, D.B., Robinson, C.J. (eds.) Computational Intelligence, The Expert Speak. IEEE Press, Piscataway (2003)
27. Silva, C.A., Sousa, J.M.C., Runkler, T.A.: Optimization of logistics systems using fuzzy weighted aggregation. Fuzzy Sets Syst. **158**, 1947–1960 (2007)
28. Silva, C.A., Sousa, J.M.C., Runkler, T.A.: Rescheduling and optimization of logistics processes using GA and ACO. Eng. Appl. Artif. Intell. **21**, 343–352 (2008)
29. Selim, H., Ozkarahan, I.: A supply chain distribution network design model: An interactive fuzzy goal programming-based solution approach. Int. J. Adv. Manuf. Technol. **36**, 401–418 (2008)
30. Fink, A., Rothlauf, F. (eds.): Advances in Computational Intelligence in Transport, Logistics, and Supply Chain Management. SCI 2009, vol. 144. Springer, Heidelberg (2009)
31. Awasthi, A., Chauhan, S.S., Goyal, S.K.: A multi-criteria decision making approach for location planning for urban distribution centers under uncertainty. Math. Comput. Model. **53**, 98–109 (2012)
32. Sanders, N.R.: Big Data Driven Supply Chain Management - A Framework for Implementing Analytics and Turning Information to Intelligence. Person Education, Upper Saddle River (2014)
33. McKinsey Global Institute: Big Data: The next frontier for innovation, competition and productivity (2011)
34. Peter, M., Timothy, G.: The NIST Report, Definition of Cloud Computing (2009)
35. Zhang, Q., Cheng, L., Boutaba, R.: Cloud computing: state-of-the-art and research challenges. J. Internet Serv. Appl. **1**(1), 7–18 (2010)
36. Grace, L.: Basics about Cloud Computing, Software Engineering Institute, Carnegie Mellon University, USA (2012). http://www.sei.cmu.edu/library/assets/whitepapers/Cloud computingbasics.pdf. Accessed August 2013
37. Mell, P., Grance, T.: The NIST definition of cloud computing v15. Version 15 (2009). http://csrc.nist.gov/groups/SNS/cloud-computing/cloud-def-v15.doc. Accessed August 2013
38. Assunção, M.D., Calheiros, R.N., Bianchi, S. Netto, M., Buyya, R.: Big Data computing and clouds: Trends and future directions. J. Parallel Distrib. Comput. (2014). (in press, corrected proof, available online)
39. Sabbaghi, A., Vaidyanathan, G.: Effectiveness and Efficiency of RFID technology in Supply Chain Management: Strategic values and Challenges. J. Theor. Appl. Electron. Commer. Res. **3**(2), 71–81 (2008)

Making Sense of Governmental Activities Over Social Media: A Data-Driven Approach

Brunella Caroleo[✉], Andrea Tosatto, and Michele Osella

Istituto Superiore Mario Boella (ISMB), Via P.C. Boggio, 61, 10138 Turin, Italy
{caroleo,tosatto,osella}@ismb.it

Abstract. Although social media attracted significant interest from governments throughout the globe, the challenge of a successful exploitation of big social data to gain valuable insights in the decision making process is still unmet. This paper aims to provide policy makers with hints and actionable guidelines for a data-driven analysis of the social accounts they manage. To this aim, we firstly propose a three-dimensional modular framework to structure the analysis; then, the logical steps required within this framework for meaningfully process big social data are detailed by suggesting text mining techniques useful for the analysis. The proposed data-driven approach could lead public administrators to a better understanding of their use of social accounts and to measure the community engagement around some topics of interest. Findings can constitute fresh insights from which public policy makers may draw for enhancing the community involvement and for becoming far more reactive to the citizenry's needs.

Keywords: Big data · Social media · Social data · Data-driven decision making · Policy making · Text mining

1 Introduction

Looking at the burgeoning big data landscape [1–4], it is immediate to realize that one of the key underlying pillars has its roots in the social media realm. In fact, social platforms – whose advent and relentless rise have marked a revolution in the last decade [5, 6] – make available unprecedented volumes of information at a prodigious rate. As a result, the undisputable game-changing role played by social media in providing a window on society has attracted the attention of forward-looking public (and private as well) decision makers.

Although social media may be seen as a goldmine of opportunities, governments are only starting to tap the vast potential unleashed by the social data deluge. Prominent barriers at work – for which the reader is referred to sections two and three – have to do with the absence of a universal approach to the big social data analysis: customized metrics and methodologies should be adopted according to different objectives. For this reason, the 'black-box' approach typically adopted by off-the-shelf social-insight tools does not meet the needs of data-driven policy making.

© Springer International Publishing Switzerland 2015
B. Delibašić et al. (Eds.): ICDSST 2015, LNBIP 216, pp. 34–45, 2015.
DOI: 10.1007/978-3-319-18533-0_4

Taking stock of the array of obstacles still standing on the way to full-fledged data-driven decision making in the public sector, the contribution of the present paper may be framed around two main axes:

- *Measurement*, related to the evidence-based appraisal of magnitude, patterns, performances and impacts of public bodies' social media presence.
- *Comprehension*, concerning the transformation of collected qualitative/quantitative measurements into actionable insights, guidelines and lessons learnt on which policy makers may rely for targeting underserved needs, shaping effective policy actions and rethinking public intervention.

In connection with the two afore-mentioned dimensions, the aim of this paper is to present a brand-new conceptual framework – accompanied by a sequence of logical steps that schematize the 'pipeline' required for meaningfully process sheer volumes of big social data – which aspires to become the everyday working tool of policy makers committed to leverage social media as springboard for unleashing a new wave of social-enhanced data-driven decision making.

The paper is structured into five sections. Section two provides a multidisciplinary theoretical background to the presented work. Section three illustrates pivotal principles underlying the proposed framework. Section four hones in on technologies, methods and tools required by the public sector (and not only) to turn data-driven social-enhanced decision making into reality. Finally, section five provides some conclusive remarks as well as some directions for stimulating future research.

2   State of the Art

In a recently published article [8], Data-Driven Decision Making (D3M) is defined as "the practice of basing decisions on the analysis of data rather than purely on intuition"; Google experts in data-driven innovation state that "data for decision making means using both real-time and historical data to inform decisions in the present" [9]. In general, D3M is considered in the literature as a 'style' of decision making that heavily involves data in the decision process to achieve stakeholder goals. A deep review about D3M reveals that its domains of application include business, medicine, education and government [10–12], with the aim of collecting and analysing data to make work or services more effective. The concept of using data for more informed D3M relies on the bigger 'container' of data-driven innovation [13], which accounts both for making decisions and improving efficiency. Literature unanimously agrees in considering data of great value for decision making as a catalyst of innovation, and scholars often tried to quantify such value, starting from the evidence of a positive relationship between data and economic growth [14, 15]. Obviously data alone cannot improve performance: in every setting the decision maker should combine data-driven information with its human experience and intuition to decide the best course of action. Furthermore, data are meaningless if not compared to other data, visualized in context or analysed for significance. Summing up, "data do not objectively guide decisions on their own - people do" [16]: the decision maker must work tied to data to produce ground-breaking innovations and valuable insights.

There is no shortage of examples of how data could efficiently be used in the private sector [14]: it is enough to mention big companies like Google, Facebook and LinkedIn that are using data to guide strategic business decisions. The data-driven process improvement is also a great opportunity for the public sector [17, 18]. The most impressive and popular example of D3M in the public sector is CompStat [19, 20], the data-driven management model adopted since 1994 by the New York City Police Department to improve crime reduction and quickly diffused across the United States. Other relevant examples of D3M in the public sector could be the application of advanced analytics to ease fraud detection in governmental organizations [21, 22], or the engagement of citizens in public matters via social media with the aim of improving and democratizing the policy making process [23].

Two aspects arise while reviewing the literature about D3M in the policy making process: on the one hand, although the public sector is one of the economy's most data-intensive sectors, scholars agree in affirming that it does not exploit the full potential of the data it generates or the potential of data generated elsewhere [15, 22, 24]. On the other hand, in most countries policy making has traditionally been a static, top-down process with citizenry having only a passive role [25]: in this sense, the emerging Web 2.0 technologies have dissolved many technical barriers to widespread citizen engagement. As Osella states [5], it is desirable that governments make a step towards citizens rather than expecting the citizenry to move their content production activity onto the 'official' spaces created for e-participation [26].

Current research on the use of big social data in government reveals the need for specific methodological approaches and for specific tools, which differ from those used hitherto in e-government studies [6]: e.g. the peculiarity of massive social data ushers-in new opportunities to enhance data visualization, study unstructured data, or develop sentiment analysis [27]. Also the European Commission cares about the use of big social data for better D3M, as evidenced by the number of projects funded under the FP7 umbrella (e.g. UniteEurope, PADGETS, SENSE4US[1]).

Even though over the last years there has been a growing adoption of social media platforms by government bodies [6, 28, 29], the challenge of a successful implementation of D3M using big social data over the whole spectrum of public decision making is still unmet [30]. The low uptake of D3M enhanced by big social data in the public sector could be due to the fact that often D3M methods are only available to skilled professionals: their techniques frequently leverage sophisticated models and rely on domain expertise [31]. Moreover, such methods typically require manually-entered data or a proper tuning of parameters, making them difficult to use and error-prone [31]. This results in developing tools for public administrations that are not highly customizable (black-box alike) and therefore do not allow a personalized investigation of their social channels. As a result, most of currently available social media tools are limited to the elaboration of a profile analysis that allows scrutinizing a certain profile while returning some standardized metrics. Even though many of these tools offer the benchmarking of

---

[1] www.uniteeurope.org, www.padgets.eu, www.sense4us.eu (visited on 02/23/2015).

a profile page versus similar ones, policy makers are left with a set of social metrics and are not able to fully comprehend the use of their social media channels and the missing potentialities arising from big social data challenges.

3 Framework

The main gap emerging from the literature review is that public administrators are actually not able to use big social data to gain valuable insights for rethinking governance. As discussed in the previous section, most of currently available social media tools elaborate profile analyses meant to better understand the use of public bodies' social channels and their positioning with respect to similar entities. The proposed approach begins with a similar investigation (the 'megaphone' dimension) and then suggests how to overcome the limits of a typical profile investigation accounting for higher levels of analysis (Fig. 1). We believe that the suggested framework could place at policy makers' fingertips useful guidelines to structure the analysis of their social accounts by outlining on the one hand the logical steps that should be followed (detailed in Sect. 4) and, on the other, by distilling precious hints for tailoring the social accounts investigation.

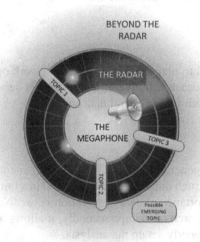

Fig. 1. Graphic representation of the proposed conceptual framework.

The proposed approach focuses on the social networks in which the text is the predominant language of communication, i.e. Twitter and Facebook. The framework includes three modular dimensions, corresponding to increasingly sophisticated levels of understanding of the entity's social accounts: in this way, policy makers could decide how to fit, size and manage their analysis.

The first dimension, labelled 'the Megaphone', is aimed to describe how a social entity is positioned in the galaxy of social media. This level, ideally similar to a megaphone that amplifies the voice of the social entity, results in a detailed investigation of the contents published by the public body on the official social network pages. The deriving profile and positioning analysis could be based on a number of cross-platform metrics, and returns the most relevant topics the social entity speaks about on its official pages, together with the most peculiar keywords used for each topic and some information on the entity's communication patterns.

In the second dimension, 'the Radar' should probe the activities of the community around each topic identified in the megaphone phase and measure how and how much the community is involved on it. As a radar finds the position and the velocity of objects in a certain range, the proposed 'social radar' could be used to detect, and even measure, the reactions of the community connected to the social entity. Moreover, the mood of the community around each topic could be investigated, together with the keywords of the discussions appearing in the radar screen of the public body.

Finally, the framework includes the dimension 'Beyond the Radar', whose aim is to describe the echo generated in the open Web around topics of interest for the social entity. This phase will extend to the open social Web the considerations made in the previous step in order to map beyond the radar range the positioning of public opinion with respect to the topics identified in 'the Megaphone' dimension. The resulting benchmarking could provide the social entity with useful hints and guidelines for the improvement of its social contents and communication.

4   Logical Steps

The logical steps required to make the proposed framework operational are presented in Fig. 2. In the following, each activity of the flowchart is detailed with the purpose of giving to the reader a toolkit for big social data analysis.

4.1   Data Observation

Although big data could generally be hard to study, big social data analysis can take advantage of the fact that such data could be consumed in a human-readable format via social networks. The importance of data observation is twofold: first of all, it is useful to test the preliminary research questions; secondly, it allows to meter the entity of big social data allowing to properly set up the analysis.

4.2   Data Fetching

In order to be analysed, social data should be dumped from providers' data warehouses to the analyst infrastructure. This operation could be unexpectedly complicated: the lack of a standard social data format requires the usage of *ad hoc* techniques to manipulate data coming from each provider [32]. For this reason, the main task of the fetching step is to harmonize data in order to work with a uniform format. The more the analysis requires different social data sources, the more the harmonization of different data sets becomes a tricky task. Many works in the literature address this problem: as an example, in [26] Charalabidis et al. present a meta model that faces this challenge abstracting the

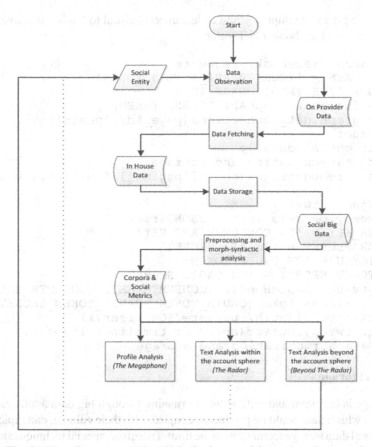

Fig. 2. Flowchart representing the logical steps of the proposed approach.

social entities in a common data interface. For the sake of the paper, as we explained in the previous sections, we propose a simple yet efficient model to store and harmonize Twitter and Facebook data. The main idea is that, albeit with some approximations, Facebook "posts" and Twitter "statuses" could be considered the same meta-entity. Moreover, Facebook "likes" are comparable to Twitter "favourites", Facebook "shares" are quite the same actions of Twitter "retweets" and, finally, Facebook "comments" and Twitter "replies" could almost be treated in the same way.

The data harmonization could therefore be implemented as:

```
{
    "id": "twitter or facebook id"
    "status" : "status or tweet",
    "likes": "number of like or stars",
    "shares": "number of share of retweet",
    "replies": [ "array of replies or comments" ],
    "source": "twitter or facebook",
    "creation_time": "creation time"
}
```

Below we give two simple Python code snippets helpful to fetch and convert social data coming from Facebook and Twitter.

```
import json, facebook, requests
ACCESS_TOKEN = 'YOUR_FB_API_ACCESS_TOKEN'
page_id = 'THE_TARGET_PAGE_ID'
graph = facebook.GraphAPI(ACCESS_TOKEN)
posts = graph.get_connections(page_id, 'posts')
while True:
for d in posts['data']:
  # Data harmonization and storage
  posts = requests.get(posts['paging']['next']).json()

import json, tweepy
from tweepy.parsers import JSONParser
CONSUMER_KEY = 'TW_CONSUMER_API_KEY'
CONSUMER_SECRET = 'TW_API_SECRET'
OAUTH_TOKEN = 'TW_OAUTH_TOKEN'
OAUTH_TOKEN_SECRET = 'TW_OAUTH_SECRET'
auth = tweepy.OAuthHandler(CONSUMER_KEY, CONSUMER_SECRET)
auth.set_access_token(OAUTH_TOKEN, OAUTH_TOKEN_SECRET)
api = tweepy.API(auth, parser=JSONParser())
for s in tweepy.Cursor(api.user_timeline).items():
    # Data harmonization and storage
```

4.3 Data Storage

Data storage is common and critical thread running through big data applications: the velocity on which data could be produced, coupled with their volume, can rapidly push the traditional data storage techniques to the limit. Therefore, special techniques and tools are needed to handle big data. Apropos of this, it appears uncontroversial that the standard de facto of big data storage is the Hadoop Distributed File System (HDFS) [33, 34], a distributed file system designed to run on commodity hardware. The main idea behind HDFS is "moving computation is cheaper than moving data": a computation requested by an application is much more efficient if it is executed close to the data it operates on. This is especially true when the size of the data set is huge.

Many other tools could be used to accomplish this task. As an example, we could cite columnar storage engines like Cassandra, or NoSQL databases like MongoDB. For the sake of the subsequent analysis, data are assumed to be queried so that portions of the original data set could be extracted.

4.4 Preprocessing and Morph-Syntactic Analysis

Once the data are stored, for ease of analysis it is necessary to convert the collected text documents (in unstructured form) to a vector representation (a structured form): parsing is the first step in this conversion. From now on, text analysis bases on at least three complementary skills in close symbiosis between them, i.e. linguistics, computer science and statistics. There are two broad-based approaches to analysing *corpora*. The first is the *bag-of-words* method, whose basic underpinning is counting words in the text, plus understanding how these words are syntactically (structurally) related to each other. The second approach, which is more *linguistic-oriented*, posits that to truly understand *corpora*, you have to move beyond syntax (structure) to semantics (meaning of words). Different tools are used in the two approaches: statistical environments like R, SAS or SPSS are typically chosen in the bag-of-words approach, while in the linguistic-oriented branch the statistical analyses are preceded by examinations in other environments, like the Stanford Parser and TreeTagger for the English language, or the TaLTaC software for Italian corpora.

In order to apply either approach, the text document is firstly parsed to find the autonomous units of analysis (e.g. words) contained in it. This process generally involves a preprocessing phase and a morph-syntactic analysis. As in every data analysis, the preprocessing phase is impossible to standardize since it depends on the research objectives. In general, this phase envisages the following tasks:

- Sentence segmentation and tokenization: sentence boundary identification is challenging because punctuation marks are often ambiguous. At the end of this level of analysis, the *corpus* is split into text segments, i.e. tokens.
- Normalization: standardization of spellings in the text (accents/apostrophes, lowercase/uppercase, abbreviations), homograph disambiguation, identification of named entities (proper names, place names, companies, people, newspapers), identification of numbers (time, dates, currencies).

The morph-syntactic analysis, that typically follows the preprocessing of data in the linguistic-oriented approach, generally includes:

- Grammatical tagging, also known as Part-of-Speech (POS) tagging or word-category disambiguation: it is the process of assigning word classes (lexical categories) to each token. Besides being useful for word sense disambiguation, giving a unique tag to each word reduces the number of parses. Possible methods for POS tagging are rule-based POS tagging, transformation-based tagging, stochastic (probabilistic) tagging.
- Lemmatization and stemming: they have the same goal of reducing inflected or derived words to their stem or root forms, but they are slightly different in their approach. Stemming follows a crude heuristic process that chops off the ends of the words. Lemmatization uses a standard vocabulary and morphological analysis of words for reducing an inflectional form of the word to its base or dictionary forms, i.e. lemma.
- Identification of structures/recognition of multiwords: not only simple words but also complex words, which are syntagmatic combinations of terms, contribute to specific concepts definitions. It is, in fact, common to find sequences of words that are semantically linked and co-occurring regularly, which make them autonomous units of analysis.

4.5 Profile Analysis/Positioning: 'the Megaphone'

The megaphone phase consists in understanding how the public body is 'spreading' its voice on its social network pages, focusing only on the public body publications, and not on the interactions with the community. This is achieved through a profile and positioning analysis of the social entity in the social media constellation: this stage takes stock of the communication pattern adopted by the social entity and, thus, is able to detect its main topics of discussion.

This task firstly requires the computation of some descriptive statistics of the social entity's accounts (e.g. the frequency of posts' publication, the number of mentions and retweets, frequency of URLs/images/videos publication). The investigation also includes the analysis of language used (e.g. local language/English, legal/popular) and the detection of the most frequent keywords (e.g. hashtags) used in the social entity's pages.

Then, in order to identify the main topics the social entity speaks about on its social pages, a more detailed text analysis could be performed. This action requires the creation of the term-document matrix (or the transposed document-term matrix), that is the starting point for any text mining activity. Since big social data have sufficiently large volumes from the statistical point of view, a multidimensional statistical analysis could be performed [35], aimed for example to provide representations of the phrase-level text, or to measure the association between words, the strength of their ties and their contexts of use. Note that text processing could be extremely diversified, as it is strictly linked to the objectives to be achieved and to the strategy of processing and analysis chosen. As an example, the multidimensional mapping deriving from a correspondence analysis could untangle the *corpora* complexity, leading to fewer factors that mainly characterize the *corpus*. Resulting factors (semantic axes) could be interpreted thanks to a simultaneous graphical representation of words and texts in the same plot: the cross-reading of the results by analysts will lead both to the selection of the most relevant topics of discussion in the social entity's accounts and of the peculiar words used in each topic.

Furthermore, the analysis of word co-occurrences, the analysis of concordances and text clustering techniques could allow to measure the association between words and their contexts of use, leading to the reconstruction of 'models of meaning' that will give useful information on the most discussed topics, also providing details on the communication patterns peculiar of the social entity.[2]

4.6 Text Analysis Within the Account Sphere: 'the Radar'

Once the most discussed topics has been identified, the radar should 'probe' the activities of the community around each topic and measure how and how much the community is involved on it.

After the computation of some descriptive statistics pertaining to the social entity's community (e.g. the number of followers/fans, their characterization, most popular comments in terms of likes/retweets), the text analyses proposed in the previous step could be replicated here limiting to the community's posts/tweets. That is, the text analyses will be performed only on the Facebook comments to the social entity's posts or, in the case of Twitter, to tweet replies or to tweets containing the entity as mentions. In this way, it is possible to investigate how the community reacts to the social entity's posts and tweets and their level of engagement.

---

[2] Due to space limits, it has not been possible to enter here code snippets for these text analyses. For further details and to access sample codes the reader is referred, for example, to the public repository http://www.rdatamining.com (visited on 02/23/2015).

Community interactions can be clustered around the topics identified in the previous step and further analysed, firstly measuring the interest of the community towards a selected topic and its keywords, which do not necessarily coincide with the social entity's ones. Then, sentiment analysis could provide relevant details on the possible mood associated to these topics in the community. Furthermore, the analysis of word co-occurrences and concordances could allow to measure the association between words and their contexts of use, enriching semantic relationships among terms.

4.7 Text Analysis Beyond the Account Sphere: 'Beyond the Radar'

After the radar probed the social entity's community engagement, the investigation proceeds beyond the radar, or rather outside the limited circles of the public body's community: the open Web. In this step the study investigates how each topic is addressed in the social Web, how the most peculiar keywords (detected in the previous steps) are dealt with, how and when they are used. Word co-occurrences analysis could be very useful in this phase, since it makes possible to detect some relevant keywords that are very common outside the social entity's radar, even if they are not frequently used by its own community. Data extraction could be performed via hashtag queries for Twitter data, or spidering the feed APIs for Facebook content.

At this juncture, the analysis results in a positioning of the social entity around its most relevant topics of discussion, its peculiar keywords and its communication patterns, providing hints and guidelines for the improvement of social contents and communication.

5 Conclusions

Coming to conclusions, the paper unveils a brand-new conceptual framework meant to support public decision makers in leveraging the abundance of social media data for policy purposes. The proposed framework has its roots in intensive fieldwork activities conducted while supporting a cohort of policy makers with whom authors collaborate on a regular basis. Although natively conceived as working tool for the public sector, the framework may be potentially adapted for private sector usage thanks to the inherent generalizability of its founding pillars; that said, this opportunity for broadening the scope of target users is out of the paper scope.

Casting a light on the public sector, the present essay encourages forward-looking policy makers to dip their toes into the challenging but rewarding waters of data-driven decision making by looking at the two deeply intertwined dimensions defined at the outset of the paper: measurement – in the guise of evidence-based appraisal of metrics stemming from social media activities – and comprehension, i.e., the transformation of collected qualitative/quantitative measurements into fresh insights from which to draw for enhancing the community involvement and for becoming more reactive to the citizenry's needs.

Among the numerous elements of novelty brought by the data-driven approach, the prominent ones have to do with the nature of the conceptual framework underpinning the approach. Its modular shape, in fact, fits with a progressive awareness and understanding of social media mechanisms: consequently, on the contrary to traditional 'black-box' tools, more degrees of freedom are left to policy makers, who can customize the analysis according to their needs, priorities and skills. In addition, its sequential arrangement mitigates barriers ingrained on the civil servants' side in terms of acquaintance with data thanks to a sequence of 'bite-sized' logical steps that schematize the 'pipeline' through which massive and machine-readable social data can morph into enhanced policy intelligence inputs in a cost effective way.

Furthermore, it is important to discuss also some of the limitations that characterize the presented findings, as they may represent an interesting starting point for future research. As hinted at in the description of logical steps, manifold phases of the analysis depend on needs and priorities of policy makers, which may vary from time to time. As a result, it goes without saying that activities and algorithms being part of each logical step cannot be thoroughly standardized by means of a fully-automated procedure. Finally, although the proposed approach unravels to a significant extent the complexity of data science in the social media realm, the uneven data culture and data literacy among policy makers still represent a hurdle to acceptance and adoption of a practice in this vein.

References

1. McKinsey Global Institute: Big data: The next frontier for innovation, competition, and productivity, Washington DC (2011)
2. Snijders, C., Matzat, U., Reips, U.D.: "Big Data": big gaps of knowledge in the field of internet science. Int. J. Internet Sci. **7**(1), 1–5 (2012)
3. Laney, D.: 3-D Data Management: Controlling Data Volume, Velocity and Variety. Gartner Report (2001)
4. Gerard, G., Haas, M.R., Pentland, A.: Big data and management. Acad. Manag. J. **57**(2), 321–326 (2014)
5. Osella, M.: A Multi-Dimensional Approach for Framing Crowdsourcing Archetypes. Ph.D. thesis (2014). http://porto.polito.it/2535900/. Accessed 23 February 2015
6. Criado, J.I., Sandoval-Almazan, R., Gil-Garcia, J.R.: Government innovation through social media. Gov. Inf. Q. **30**(4), 319–326 (2013)
7. Bollier, D.: The Promise and Peril of Big Data. The Aspen Institute, Communications and Society Program, Washington (2010)
8. Provost, F., Fawcett, T.: Data science and its relationship to big data and data-driven decision making. Big Data **1**(1), 51–59 (2013)
9. Hemerly, J.: Public Policy Considerations for Data-Driven Innovation. Computer **46**(6), 25–31 (2013)
10. Marsh, J.A., Pane, J.F., Hamilton, L.S.: Making Sense of Data-Driven Decision Making in Education. Rand, Santa Monica (2006)
11. McAfee, A., Brynjolfsson, E.: Big data: the management revolution. Harvard Bus. Rev. **90**(10), 60–66 (2012)
12. Sackett, D.L.: Evidence-based medicine. Semin. Perinatol. **21**, 3–5 (1997)
13. U.S. Chamber of Commerce Foundation: The future of data-driven innovation (2014)
14. Brynjolfsson, E., Hitt, L.M., Kim, H.H.: Strength in Numbers: How Does Data-Driven Decisionmaking Affect Firm Performance? (2011). http://ssrn.com/abstract=1819486. Accessed 23 February 2015

15. Manyika, J., Chui, M., Brown, B., Bughin, J., Dobbs, R., Roxburgh, C., Byers, A.H.: Big data: The Next Frontier for Innovation, Competition, and Productivity. McKinsey Global Institute (2011)
16. Spillane, J.P.: Data in practice: conceptualizing the data-based decision-making phenomena. Am. J. Educ. **118**(2), 113–141 (2012)
17. Desouza, K.C., Jacob, B.: Big Data in the Public Sector: Lessons for Practitioners and Scholars. Administration & Society. November 6, 2014. doi:10.1177/0095399714555751
18. Kim, G.-H., Trimi, S., Chung, J.-H.: Big-data applications in the government sector. Commun. ACM **57**(3), 78–85 (2014)
19. McDonald, P.P.: Managing police operations: Implementing the New York Crime Control Model – CompStat. Wadsworth Publisher, Belmont (2002)
20. Godown, J.: The CompStat process: Four principles for managing crime reduction. Police Chief. **76**(8), 36–42 (2009)
21. Price-Waterhouse Coopers: Fighting Fraud in the Public Sector. In: Global Economic Crime Survey (2011). https://www.pwc.com/en_GX/gx/psrc/pdf/fighting_fraud_in_the_public_sector_june2011.pdf. Accessed 23 February 2015
22. Cebr, Centre for Economics and Business Research Ltd. Data equity - Unlocking the value of big data. Report for SAS (2012)
23. Ferro, E., Loukis, E., Charalabidis, Y., Osella, M.: Policy Making 2.0: from theory to practice. Gov. Inf. Q. **30**(4), 359–368 (2013)
24. OECD: Data-driven Innovation for Growth and Well-being (2014)
25. Tapscott, D., Williams, A.D., Herman, D.: Government 2.0: Transforming Government and Governance for the Twenty-First Century. New Paradigm, Toronto (2008)
26. Charalabidis, Y., Gionis, G., Ferro, E., Loukis, E.: Towards a systematic exploitation of web 2.0 and simulation modeling tools in public policy process. In: Tambouris, E., Macintosh, A., Glassey, O. (eds.) ePart 2010. LNCS, vol. 6229, pp. 1–12. Springer, Heidelberg (2010)
27. Chun, S.A., Reyes, L., Luis, F.: Social media in government. Gov. Inf. Q. **29**(4), 441–445 (2012)
28. Bertot, J.C., Jaeger, P.T., Munson, S., Glaisyer, T.: Social media technology and government transparency. Computer **43**(11), 53–59 (2010)
29. Snead, J.T.: Social media use in the u.s. executive branch. Gov. Inf. Q. **30**(1), 56–63 (2013)
30. Benčina, J.: Web-based decision support system for the public sector comprising linguistic variables. Informatica **31**(3), 311–323 (2007)
31. Duggan, J.: The case for personal data-driven decision making. In: Proceedings of the 40th International Conference on Very Large Data Bases, 7(11) (2014)
32. Kadadi, A.: Challenges of data integration and interoperability in big data. In: IEEE International Conference on Big Data, 27–30 October 2014, pp. 38–40 (2014)
33. Shafer, J., Rixner, S., Cox, A.L.: The hadoop distributed filesystem: balancing portability and performance. In: Proceedings of IEEE International Symposium on Performance Analysis of Systems & Software (ISPASS), pp. 122–133 (2010)
34. Pandey, S.: Prominence of mapreduce in big data processing. In: Fourth International Conference on Communication Systems and Network Technologies (CSNT), 7–9 April 2014, pp. 555–560. IEEE (2014)
35. Bolasco, S., Chiari, I., Giuliano, L.: Statistical analysis of textual data. In: Proceedings of the 10th International Conference JADT, LED, Milano, vol. 2, p. 1330 (2010)

Data-Mining and Expert Models for Predicting Injury Risk in Ski Resorts

Marko Bohanec[1(✉)] and Boris Delibašić[2]

[1] Department of Knowledge Technologies, Jožef Stefan Institute,
Jamova cesta 39, 1000 Ljubljana, Slovenia
marko.bohanec@ijs.si
[2] Faculty of Organizational Sciences, University of Belgrade,
Jove Ilica 154, 11000 Belgrade, Serbia
boris.delibasic@fon.bg.ac.rs

Abstract. This paper proposes several models for predicting global daily injury risk in ski resorts. There are three types of models proposed: based on data mining, expert modelling, and a combination of both. We show that the expert model that represents the judgment of injury risk experts in the analyzed ski resort is 10–15 % less accurate than data mining models. We also show that expert models refined with data-driven analysis can produce models that are in line with accuracy as data mining models, but in addition show some advantages, like transparency, consistency and completeness.

Keywords: Decision analysis · Data mining · Expert modelling · Multi-attribute models · DEX · Orange · Ski injury

1 Introduction

In this paper we propose several models for predicting global daily injury risk in ski resorts. Most ski resorts are equipped with ski lift tickets entrance systems which collect massive data from skiers' movement through ski lift gates on a ski resort. In this paper we show how this big data can be used for injury prediction, which is a crucial activity in ski resorts. On the ski resort there are various types of data which can help in injury prediction, such as meteorological data and ski lift entrance data. This data is big in terms of the "3V": high volume (all ski lift entrances are recorded), high velocity (there is a constant throughput of skiers in the system), and high variety (in terms that an extensive pre-processing is required to integrate all data in an adequate form for the analysis).

Ski injuries are rare events occurring at small rates at mountains (less than 0.2 % per total skier days in a ski resort [1]). However, due to the large number of alpine snowsport participants, the number of ski injuries is still substantial. The number of injuries is not directly correlated to the number of skiers in the ski resort each day, so predicting the number of injuries on a daily level is hard.

Ski resort managements are usually interested in predicting whether ski injuries will occur on a certain day or time period as well as on how many ski injuries will occur because rescue staff and rescue equipment can be better allocated, and ski injuries

© Springer International Publishing Switzerland 2015
B. Delibašić et al. (Eds.): ICDSST 2015, LNBIP 216, pp. 46–60, 2015.
DOI: 10.1007/978-3-319-18533-0_5

handled in a more efficient way. On the other hand, if ski resorts cannot forecast the number of ski injuries efficiently, they are at risk of allocating too few resources and not handling injuries efficiently, which could have many negative impacts on the ski resorts, such as legal issues with the clients, making a negative image of the ski resort and losing trust with its clients.

Predicting the risk of injury/accidents is an extremely interesting research topic in many disciplines, for instance in occupational accident analysis [2], road traffic analysis [3, 4], and sports [5]. Predicting the risk of ski injuries, however, is still missing in the literature. There are at last two reasons for this: (1) Ski lift transportation data is often unavailable, (2) ski injury research as well as ski resort management is still very traditional, so big data and data mining methods are currently underexploited. Skiing and snowboarding have become staples in winter recreation, accounting for 58 million visits to US ski resorts in 2005 [6]. On the other hand, ski injuries have been identified as the most common cause of sports-related injuries [7]. Although the rate of ski injuries has reached a stable rate more than 10 years ago and is at the level from 0.2 % to 0.3 % [8, 9] reports that 5700 (from 700,000) severe injured skiers needed medical attention, and the cost of this was considerable. There are papers that predict the number of skier days in a ski resort during the whole season [10], which could give some hint on the number of expected injuries during the whole season as injury rates are correlated with the number of skier days.

In general, predicting injury risk on a daily basis is difficult, as it can depend on many factors, such as congestion on ski slopes, meteorological conditions, snow conditions, structure of skiers on the mountain (number of beginners, advanced skiers, children).

Ultimately, we wish to develop models and software tools that would help the management of the ski resort Mt. Kopaonik at assessing dangerous days and consequently allocating resources for handling ski injuries. In this paper we utilize ski lift transportation data, as well as publicly available meteorological data, to predict injury occurrence in the ski resort. Specifically, we are interested to predict daily risk of injury to support making informed decisions. We applied a combination of data mining and expert modelling to develop risk assessment models. Using data from the resort, we applied several machine learning algorithms (Classification Trees, CN2, Naïve Bayes, Logistic Regression, SVM (Support Vector Machine) and k-Nearest Neighbours). Using the results of data analysis and consulting a domain expert, ski resort safety manager, we employed a qualitative modelling method DEX, developed a multi-attribute model, and assessed its performance.

2 Data and Methods

The data used in this research is from the largest Serbian ski resort Mt. Kopaonik. This ski resort has a total ski lift capacity of 32,000 ski lift transportations per hour, which makes it one of the largest ski resorts in South-East Europe. The data spans over six consecutive years (2005 to 2010) and includes 605 skiing days. Regulations on Mt. Kopaonik are that each person must buy a ski pass in order to use ski lifts. The RFID ski pass is used each time a person wants to enter a ski lift through a ski lift gate.

Therefore, for all skiers (in this paper skiers will be used as a generic term for all persons using ski lift gates to enter ski lifts, i.e., skiers, snowboarders, etc.), motion data is collected on ski lift gates and stored in the central database. Mt. Kopaonik has an injury rate of 1.5 IPTSD (injuries per thousand skier-days), which is similar to other ski resorts.

The RFID-collected data allowed us to generate statistics that we used for ski injury prediction. Attributes, which are used, or could be used, for predicting the injury risk on a specific day, were formulated based on interviews with ski safety experts from Mt. Kopaonik. The used attributes with short descriptions and basic statistics are given in Table 1.

Table 1. Basic statistics of global ski resort attributes used for analysis at the daily level.

Attribute	Description	Min	Max	Median	Mean
numSkiers	Number of skiers	15.00	33085	8639	9255.34
numPasses	Number of ski lift transportations	54.00	73378	26718	26017.05
utilization	Percentage of ski lift capacity used	0.00	0.48	0.20	0.19
avgNumRuns	Avg. skiers' number of runs	1.00	40.00	11.17	11.27
avgAvgTimeRun	Avg. skiers' passage time	11.00	66.24	25.55	25.96
avgMinTimeRun	Avg. min. skiers' passage time	3.50	48.95	11.91	12.50
avgMaxTimeRun	Avg. max. skiers' passage time	15.00	211.00	68.88	68.47
avgAvgElevation	Avg. skiers' runs elevation	113.97	302.67	209.30	207.76
avgMinElevation	Avg. skiers' runs min. elevation	36.00	248.00	126.80	131.24
avgMaxElevation	Avg. skiers' runs max. elevation	140.49	424.00	327.50	309.48
avgSpeedRun	Avg. skiers' run speed	7.00	830.00	109.71	114.34
tempAvg	Avg. temperature (Celsius)	−19.20	8.40	−2.40	−3.07
windSpeed	Wind speed (m/s)	0.50	11.40	2.90	3.31
cloudiness	Cloudiness (0-sunny, 10-clouded)	0.00	10.00	8.00	6.94
numInjuries	Number of injuries	0.00	15.00	2.00	2.58

There are in total 15 attributes in Table 1. The first 11 attributes are extracted from ski lift gates data. Notably, the number of skiers varies from 15 to more than 33,000, so predicting the riskiness of a skiing day is very important. It is not reasonable to employ the same rescuing resources in these two situations.

The analysis was carried out through the following steps:

1. data pre-processing,
2. data mining: exploring properties of the dataset and extracting injury-prediction models from data using machine learning methods,
3. developing qualitative multi-attribute risk-assessment models using an expert modelling and a combined expert-data approach.

Data Pre-processing. Data pre-processing addressed two issues with the original dataset. The first issue is related to the target attribute in the dataset, *numInjuries*, which gives the number of injuries sustained on each particular day. The purpose of this study was not to predict the exact number of injuries, but rather to assess whether ski conditions present risks for skiers. Therefore, we transformed *numInjuries* in two binary target classes:

- *areInjuries*: true if at least one injury was sustained on that day;
- *manyInjuries*: true if the number of sustained injuries was above the average.

Secondly, most of the attributes in the original dataset are numeric; even though they are perfectly suitable for data mining, we also wanted to introduce qualitative, discrete attributes, which are more suitable for expert-based development of multi-attribute models. Therefore, we transformed the original dataset (from now on called *num*, for "numeric") to additional datasets: *bin3*, in which all numeric attributes were discretized to three values: L, M, H (for "Low", "Medium", "High"), and *bin5* with five discretization values VL, L, M, H, VH (V stands for "Very"). The employed discretization principle was to produce bins which all have equal number of cases. This discretization principle showed best results compared to other basic discretization principles like mean ± standard deviation, or entropy-based binning with respect to the output variable. The subsequent data analyses involved various combinations of *num*, *bin3*, and *bin5*, as well as their union called *all*.

Data Mining. The aims of the data mining stage were twofold:

- to understand the influences of individual attributes and their interactions, and
- to develop risk-assessment models by machine learning algorithms from data.

For both tasks, we used the general-purpose machine learning and data mining tool called Orange [11, 12] (http://orange.biolab.si/).

In developing models by machine learning, our preferences were oriented to comprehensible, transparent (so-called "glass-box" or "white-box") models that would be easy to interpret and understand. Consequently, we used two main machine learning methods:

- Classification Tree, which develops models in form of decision trees [13],
- CN2, which develops models containing if-then rules [14].

However, there are other machine learning methods, which produce less transparent models, but are known to achieve high classification accuracy in a variety of data mining situations. We thus also included Naïve Bayes, Logistic Regression, SVM (Support Vector Machine) and k-Nearest Neighbours, employing their implementation in Orange.

All the learning algorithms were tested on two datasets using the schema that resembled the expected use, where the models are developed from historic data and applied on current data. Therefore, the models were developed using a learning set consisting of 490 days from the years 2005–2009, and subsequently tested on the test set of 115 days from 2010.

Developing Qualitative Multi-attribute Models. Our second approach to the development of risk-assessment models was based on qualitative multi-attribute modelling. Multi-attribute (or multi-criteria) models [15] are a tool used in decision analysis for the evaluation and selection of decision alternatives. They consist of attributes, which describe alternatives and their properties, and functions (such as utility, value of aggregation functions) that calculate alternatives' values. Multi-attribute models are usually developed in collaboration between decision makers(s), experts and decision analysts, employing their knowledge and expertise. Typically, such models are developed "manually" in the sense that all their components are defined explicitly by the developers, rather than, for example, derived from data by machine learning. This is why this approach is also referred to as *expert modelling*.

In this study, we employed the expert modelling method DEX [16]. A DEX model consists of a hierarchical structure of attributes; the hierarchy represents relations between attributes, so that higher-level attributes depend on lower-level ones. All attributes in DEX are qualitative (discrete) and their value scales consist of a set of words. The aggregation of lower-level attributes (model inputs) to higher-level ones (model outputs) is defined by decision rules, which are conveniently collected in decision tables. Within multi criteria decision analysis [17], DEX belongs to the class of qualitative multi-criteria methods, which includes methods such as MACBETH [18], DRSA (Dominance-based Rough Set Analysis [19]) and VDA (Verbal Decision Analysis [20]). The method DEX is implemented in software called DEXi (http://kt.ijs. si/MarkoBohanec/dexi.html), which facilitates the creation of DEX models and use of models for the evaluation and analysis of alternatives.

There were several good reasons for choosing DEX in this study. First and foremost, we suspected that our dataset was too small to uncover all the relevant risk factors using machine learning only, therefore we wanted to complement the machine learning approach with the expert modelling one. Second, the DEXi software provides methods that ensure the completeness and consistency of decision rules, which may provide an advantage over machine learning models, which typically do not fulfil these properties. Third, the ski-resort dataset contains groups of attributes that address different aspects of the ski resort situation, such as weather, utilization aspects, and characteristics of individual ski runs; such groups are very suitable for hierarchical treatment. Last but not least, DEX has already been used with good results in risk-assessment problems, even though in a very different area – banking [21].

Another important aspect from the expert modelling viewpoint is that the dataset, which is available in this study, provides a rare opportunity to assess the accuracy of DEX models. Even more, the results of data mining (from a learning set) can be taken into account when developing a corresponding DEX model, possibly improving the quality of its decision rules. For this reason, we actually developed two DEX models:

1. Basic DEX model: developed only through communication with the ski-management expert;
2. Enhanced DEX model: considering expert knowledge, taking into account the results of data mining (frequencies of attribute values and other properties of attributes, such as monotonicity), and employing DEXi to ensure the consistency and completeness of decision rules.

Both models were tested for accuracy on the test dataset.

3 Results

Data Mining. Results obtained from the datasets did not substantially differ between each other. Hereafter, we thus present only results based on the *bin3* dataset, as the models obtained from this dataset were the smallest and easiest to interpret.

Table 2 displays the influence of individual attributes from the *bin3* dataset on the *manyInjuries* target class using four different information-theoretic measures implemented in Orange. The aim of this analysis is to understand the strength of attributes, individually and in relation with each other. The attributes are rather clearly clustered according to their strength. The first, most influential, group of three attributes refers to the utilization and number of skiers present on the ski slopes. The second group consists of two weather attributes, *windSpeed* and *tempAvg* (excluding *cloudiness*, which seems substantially less important). The remaining least influential attributes compose the third group, which refers to the characteristics of individual ski runs.

Table 2. The strength of influence of individual *bin3* attributes on *manyInjuries*.

	Attribute	ReliefF	Inf. gain	Gain ratio	Gini
1	numPasses	0.3504	0.3103	0.1963	0.0927
2	utilization	0.2803	0.2263	0.1429	0.0655
3	numSkiers	0.2717	0.2911	0.1841	0.0868
4	windSpeed	0.1276	0.0214	0.0140	0.0070
5	tempAvg	0.1239	0.0120	0.0076	0.0039
6	avgMinTimeRun	0.1123	0.0312	0.0198	0.0102
7	avgAvgTimeRun	0.1106	0.0485	0.0307	0.0154
8	avgMinElevation	0.0864	0.0324	0.0205	0.0104
9	avgAvgElevation	0.0696	0.0265	0.0169	0.0087
10	avgMaxElevation	0.0658	0.0632	0.0402	0.0203
11	avgSpeedRun	0.0567	0.0364	0.0231	0.0122
12	avgNumRuns	0.0433	0.0303	0.0192	0.0102
13	cloudiness	0.0190	0.0240	0.0151	0.0080
14	avgMaxTimeRun	0.0096	0.0147	0.0093	0.0048

Interaction analysis was employed to identify dependencies between attributes, which proved useful in the subsequent DEX model development. Interaction gain [22] is an information-theoretic measure of irreducible dependencies between attributes:

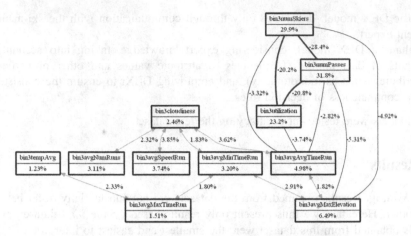

Fig. 1. Interactions between attributes in the *bin3* dataset

two attributes are said to interact when they, observed together, provide a different amount of information for predicting the target concept than when considered individually. Negative interactions indicate redundancy (i.e., attributes provide similar information), whereas positive interactions indicate synergy (i.e., two attributes together explain more than what can be estimated from each attribute individually). Figure 1 displays interaction gains between pairs of *bin3* attributes. There are two distinct attribute clusters in Fig. 1. On one hand (top right in Fig. 1), there is a strongly negatively connected group of attributes, which have already been identified in Table 2 and apparently indicate a common concept (hereafter referred to as "crowding"). The second cluster (bottom part of Fig. 1) contains attributes related to weather and skiing; positive interactions indicate that these attributes should be used together in order to predict ski injuries.

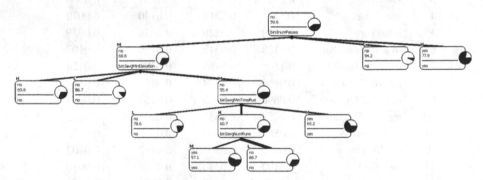

Fig. 2. Decision tree for the prediction of *manyInjuries*, developed from *bin3* by Classification Tree

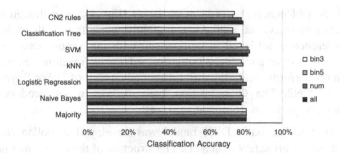

Fig. 3. Classification accuracy of machine learning models when predicting *areInjuries*

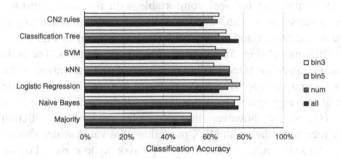

Fig. 4. Classification accuracy of machine learning models when predicting *manyInjuries*

Figure 2 presents one of the models obtained by data mining: a decision tree developed from *bin3* by the Classification Tree algorithm. The tree predicts whether *manyInjuries* occur or not. The first step of the prediction is based on the root attribute *numPasses*: when *numPasses* is=L (Low) or H (High), then *manyInjuries* are no or yes, respectively. The case *numPasses*=M (Medium) leads to a subtree that can be interpreted in a similar way. Notice that the decision tree is incomplete, because it does not provide a classification when *avgNumRuns*=H.

What is the quality of models developed by machine learning? Figures 3 and 4 show classification accuracy achieved by different algorithms and different sets of attributes on the test set when predicting *areInjuries* and *manyInjuries*, respectively. The method name Majority refers to the apriori accuracy, determined from the majority class values as follows:

- for *areInjuries*: the test set contains 94 of 115 examples where *areInjuries*=yes, so the apriori accuracy is 94/115 = 81.74 %;
- for *manyInjuries*: the majority class value is no, and the apriori accuracy is 61/115 = 53.04 %.

The prediction of *areInjuries* (Fig. 3) seems very hard, as the models do not achieve even the apriori accuracy. The only exception is SVM, which performed only marginally better, achieving up to 83.48 % when using all the attributes for classification. Overall, all learning algorithms performed similarly. Apparently, the problem is caused by a very high frequency of days when at least one injury occurs at the ski resort.

The prediction of *manyInjuries* is much better (Fig. 4), even though the classification accuracies of individual methods are slightly worse than with *areInjuries*. The substantial difference is achieved in comparison with the apriori accuracy, which is much lower in this case and easily outperformed by all machine learning models. Overall, methods perform similarly, with a slight advantage of Naïve Bayes and slight disadvantage of CN2. The Classification Tree method, which produces desirable transparent models, performs comparatively well.

Basic DEX Model. The basic DEX model was developed in collaboration with a domain expert (ski resort safety manager). The structure of the model was proposed by the analyst, whereas all decision rules were formulated by the expert. The expert was not aware of the results of data mining analysis, thus the rules reflect only his own opinion. As the expert did not feel comfortable with the skiing attributes (i.e., the attributes whose names start with "avg" in Table 1), only the remaining six attributes were included in the model and structured in the tree shown in Fig. 5a.

The basic DEX model (Fig. 5a) consists of two main subtrees. The first one, named *Crowding*, encompasses all three attributes related to the utilization of the ski slope. Similarly, the second subtree, *Weather*, groups al the weather-related attributes. The root attribute, *Skiers*, aggregates the two subtrees. All the attributes use the three-valued scale {L, M, H}. Notice, however, that the scales are ordered differently. This is according to a recommendation in DEX modelling that value scales should be ordered from "bad" to "good", in this case, from high risk to low risk. For instance, risk increases with an increasing number of skiers present on the slope, but also with decreased temperature. Hence, the corresponding H and L values appear on the left of each scale and are shown in bold typeface in Fig. 5. Similarly, values of low risk appear on the right and are shown in italics.

(a) Basic model

Attribute	Scale
Skiers	H; M; *L*
Crowding	H; M; *L*
numSkiers	H; M; *L*
numPasses	H; M; *L*
utilization	H; M; *L*
Weather	H; M; *L*
tempAvg	L; M; *H*
windSpeed	L; M; *H*
cloudiness	L; M; *H*

(b) Enhanced model

Attribute	Scale
Skiers	H; M; *L*
Crowding	H; M; *L*
numSkiers	H; M; *L*
numPasses	H; M; *L*
utilization	H; M; *L*
Weather	H; M; *L*
tempAvg	L; M; *H*
windSpeed	L; M; *H*
cloudiness	L; M; *H*
Skiing	H; M; *L*
TimeRun	H; M; *L*
avgAvgTimeRun	H; M; *L*
avgMinTimeRun	H; M; *L*
avgMaxTimeRun	H; M; *L*
Elevation	H; M; *L*
avgAvgElevation	H; M; *L*
avgMinElevation	L; M; *H*
avgMaxElevation	H; M; *L*
SpeedRun	H; M; *L*
avgNoRuns	M; L; *H*
avgSpeedRun	M; L; *H*

Fig. 5. Basic and enhanced DEX models: trees of attributes and value scales

The attribute *Skiers* appears at the root of the tree and represents the final risk assessment of the situation. In contrast with data mining, where we used two binary classes, *areInjuries* and *manyInjuries*, and in order to avoid developing a DEX model for each class, we designed *Skiers* as a three-valued attribute that assesses injury risk using the familiar scale {H, M, L}. The value L roughly corresponds to days of low risk where no injuries are normally expected (*areInjuries*=no). The value H denotes a high risk for sustaining many injuries (*manyInjuries*=yes). The value M is intermediate, denoting a situation where injuries are expected, but not many of them (*areInjuries*=yes and *manyInjuries*=no).

Based on this design of the attributes, their structure and scales, the expert defined decision rules using the computer program DEXi under supervision of the decision analyst. For each of the three aggregate attributes, *Crowding*, *Weather* and *Skiers*, the expert defined a corresponding decision table. In DEXi, the acquisition of rules takes place interactively through the so-called *elementary rules*: the user has to provide an output value for each combination of input values in the given context. For instance, the table of elementary rules for *Skiers* is shown in Table 3; all the nine combinations of values of *Crowding* and *Weather* were prepared by DEXi, and the expert provided only the values shown in the column *Skiers*. The remaining two tables in the model (in the context of *Crowding* and Weather subtrees) have 27 elementary rules each; they were defined in the same way, but are for brevity shown here using *aggregate rules* (Table 4). Aggregate rules employ symbols '*' (denotes any value), '<=' (worse than or equal), and '>=' (better than or equal), and are thus more compact than elementary rules.

The basic DEX model has advantages and disadvantages. On the positive side, it is relatively simple, transparent and comprehensible for its main creator and end user – ski safety manager. In addition to the *Skiers* attribute, which assesses the overall risk of injuries, the model defines two other attributes that also assess particular aspects of risks: *Crowding* and *Weather*. In other words, when using the model on a daily basis, the manager actually obtains three risk assessments: an overall one (*Skiers*) and two specific (with respect to *Crowding* and *Weather*). This extends the information available for making and justifying management decisions. By design, the DEX model is guaranteed to be *complete* (providing evaluations for all possible input values) and *consistent* (increased risk at each input attribute does not decrease risk at any assessed output attribute, i.e., the evaluation function is monotone). On the negative side, the basic DEX model achieves a relatively low accuracy measured on the test set: 65.22 % for predicting *areInjuries* (assumed when *Skiers*=M or H) and 66.09 % for predicting *manyInjuries* (*Skiers*=H). This is roughly 10–15 % worse than the corresponding machine learning models, a substantial difference.

Enhanced DEX Model. The enhanced DEX model was developed in order to improve the classification accuracy of the basic DEX model. The improvements were twofold: (1) including all available attributes in the model, (2) defining decision rules by taking into account the results of data analysis.

Adding the remaining skiing attributes in the model gave the tree structure shown in Fig. 5b: the "avg" attributes were collected in the new Skiing subtree and clustered into three groups: *TimeRun*, *Elevation*, and *SpeedRun*. The new attributes adopted the

Table 3. Basic DEX model: elementary decision rules for predicting risk of injury (*Skiers*)

	Crowding	Weather	Skiers
1	H	H	H
2	H	M	H
3	H	L	M
4	M	H	H
5	M	M	M
6	M	L	L
7	L	H	M
8	L	M	M
9	L	L	L

Table 4. Basic DEX model: aggregate decision rules for *Weather* (left) and *Crowding* (right)

tempAvg	windSpeed	cloudiness	Weather	numSkiers	numPasses	utilization	Crowding
L	*	*	H	H	H	<=M	H
*	L	L	H	H	*	H	H
>=M	L	>=M	M	*	H	H	H
>=M	>=M	*	L	H	*	L	M
				<=M	<=M	L	M
				*	H	L	M
				H	>=M	>=M	M
				<=M	M	>=M	M
				M	<=M	>=M	M
				>=M	H	>=M	M
				M	M	*	M
				>=M	>=M	H	M
				>=M	L	>=M	L
				L	>=M	>=M	L

same {L, M, H} value scale, ordered from high to low risk. This principle gave two strangely ordered scales {M, L, H} attached to *avgNumRuns* and *avgSpeedRun*, saying that the values M and H indicate the highest and the lowest risks, respectively. Even though this appears unusual, it is supported by data in the learning set.

Decision rules in the enhanced model were developed through three steps, which combined data analysis and expert modelling, as follows.

1. Data Analysis: We took the learning dataset and calculated conditional class frequencies for all combinations of attributes that appear in subtrees of Fig. 5b. We illustrate this process with the *Crowding* subtree (Table 5): the three columns below "Input attributes" list all the 27 combinations of the three input attributes' values, and the two columns under "Class frequencies" display the conditional frequencies of *areInjuries* and *manyInjuries*. The values N/A indicate that there were no corresponding entries in the learning dataset.

It is important to notice that Table 5 has exactly the same 27 rows as the corresponding DEX decision table. The only difference is that Table 5 provides class frequencies for each row, while a DEX decision table requires a single L, M or H value assigned instead. Hence, the aim of the following two steps is to define such value assignments.

Table 5. Combining *numSkiers*, *numPasses* and *utilization* to *Crowding*: class frequencies and elementary decision rules

	Input attributes			Class frequencies		Assigned DEX class
	numSkiers	numPasses	utilization	areInjuries=yes	manyInjuries=yes	Crowding 2 step 3
1	L	L	L	0,29	0,03	L
2	L	L	M	0,53	0,18	M
3	L	L	H	0,71	0,21	M
4	L	M	L	1,00	0,00	M
5	L	M	M	1,00	0,00	M
6	L	M	H	1,00	0,00	M
7	L	H	L	N/A	N/A	M
8	L	H	M	N/A	N/A	M
9	L	H	H	N/A	N/A	H
10	M	L	L	1,00	0,00	M
11	M	L	M	0,67	0,00	M
12	M	L	H	N/A	N/A	M
13	M	M	L	0,79	0,18	M
14	M	M	M	0,92	0,33	M
15	M	M	H	0,80	0,35	M
16	M	H	L	N/A	N/A	M
17	M	H	M	1,00	0,86	H
18	M	H	H	1,00	0,75	H
19	H	L	L	N/A	N/A	M
20	H	L	M	N/A	N/A	M
21	H	L	H	N/A	N/A	H
22	H	M	L	N/A	N/A	M
23	H	M	M	0,93	0,64	H
24	H	M	H	1,00	0,00	M→H
25	H	H	L	N/A	N/A	H
26	H	H	M	0,98	0,71	H
27	H	H	H	0,97	0,81	H

2. Assigning DEX Classes From Class Frequencies: Given the class frequencies in Table 5, and adopting the interpretation of risk scale {L, M, H} as described above, it is relatively easy to estimate a DEX class for entries with known class frequencies. Specifically, we adopted the heuristic rule:

$$Crowding = \begin{cases} L \Leftarrow p(areInjuries) < 0.5 \land p(manyInjuries) < 0.5 \\ M \Leftarrow p(areInjuries) \geq 0.5 \land p(manyInjuries) < 0.5 \\ H \Leftarrow p(areInjuries) \geq 0.5 \land p(manyInjuries) \geq 0.5 \end{cases} \quad (1)$$

Using this rule, 17 of 27 entries were assigned a DEX class as shown in Table 5 (left justified under *Crowding*). Among these, there was a notable exception at the entry 24, where the above rule assigned the value M. This assignment, however, violates the principle of dominance in comparison with entry 23, which is H. Moving from entry 23 to 24, the value of utilization changes from M to H, thus the assigned risk class should not decrease from H to M. In this and all similar cases, we enforced the principle of dominance, so that the decision tables stayed consistent. Hence, the entry 24 was assigned the value H instead of M.

3. Expert Modeling: After step 2, the decision table still contained undefined entries for which there was no evidence in the data set. For these entries, we employed a normal expert modelling process, where the values of missing entries were assigned by the expert, obeying the principles of completeness and consistency. In Table 5, such assignments are shown right justified in the *Crowding* column.

These steps gave a completely and consistently defined DEX decision table. The process was repeated for all decision tables in a DEX model (there are seven in the model shown in Fig. 5b).

Measured on the test set, the extended DEX model achieved 80.87 % accuracy for predicting *areInjuries* and 75.65 % for predicting *manyInjuries*. This is a vast improvement over the basic DEX model and is comparable with the accuracy of machine learning models. Actually, among the models that predict *manyInjuries* and use only *bin3* attributes, the extended DEX model is the second best after Naïve Bayes (78.62 %), whereas the accuracy of all the other models is below 73.91 %. This, together with the advantages indicated above, makes the enhanced DEX model an excellent candidate for implementation in the ski resort decision support system.

4   Conclusion

Supporting decisions from big data is essential for efficient management. The results of this study confirmed our initial expectations that data, collected from ski lift gates and combined with meteorological data and expertise of ski-resort managers, provide a feasible foundation for predicting injury risk in ski resorts. The study was carried out in three main stages (data analysis, machine learning and expert modelling) and produced several risk assessment models (machine learning models and qualitative multi-attribute DEX models).

In the data analysis stage, we identified some important principles that influence injury risks: strength of attributes, their ordering and interactions, and combinations of attribute values leading to high risks. These findings turned out particularly useful in the expert modelling stage.

The machine learning stage contributed several models for injury risk assessment. The models, which were developed exclusively from the Mt. Kopaonik dataset, performed similarly in terms of classification accuracy. It turned out that the concept of *areInjuries* (days with at least one injury) is hard to predict due to high frequency of injuries and associated uncertainty (injuries occurring by chance). The prediction of *manyInjuries* (days with above-the-average number of injuries) is more feasible, with achieved classification accuracy at about 80 %. Among these models, we strongly preferred those developed by Classification Tree and CN2 algorithms for their transparency and comprehensibility.

Expert modelling usually proceeds in collaboration with domain experts, but in the absence of historic data. In this study, the availability of the Mt. Kopaonik dataset provided a rare opportunity to enhance expert modelling with results of data analysis and to assess the quality of the developed models on a test dataset. We employed a combination of data mining and expert modelling approach, with which a part of the DEX model was developed using conditional class frequencies obtained from the test dataset, and the remaining part was completed through a normal DEX model-development procedure. Notably, defining this remaining part effectively extended the model to address situations for which there was no evidence in the learning dataset.

Classification accuracy of the DEX model developed in this way is about 10–15 % better than the model developed only by expert modelling, and is comparable with the accuracy of best machine learning models. Comparatively, the DEX model has some advantages: it guarantees for completeness and consistency, it is transparent and comprehensible, and provides three specific risk indicators (crowding, weather and skiing) in addition to the overall one.

Further work should also test whether making predictions on an hourly basis, rather than daily basis, would yield better predictions. Other time scales could be also relevant for testing, like weekday/weekend skiing, major holidays. The structure of skiers (number of children, elders, adults, teenagers, beginners, experts, males, females) could also play an important role in predicting injury risk. As the model was developed from data collected at only one ski resort, its generality should be tested with other ski resorts. This paper provides a solid foundation in terms of the tools and methods needed to develop such a ski risk injury assessment model.

References

1. Ruedl, G., Kopp, M., Sommersacher, R., Woldrich, T., Burtscher, M.: Factors associated with injuries occurred on slope intersections and in snow parks compared to on-slope injuries. Accid. Anal. Prev. **50**, 1221–1225 (2013)
2. Bellamy, L.J., Damen, M., Manuel, H.J., Aneziris, O., Papazoglou, I.A., Oh, J.I.: Risk Horoscopes: predicting the number and type of serious occupational accidents in the Netherlands for sectors and jobs. Reliab. Eng. Syst. Saf. **133**, 106–118 (2014)
3. Effati, M., Rajabi, M.A., Hakimpour, F., Shabani, S.: Prediction of crash severity on two-lane, two-way roads based on fuzzy classification and regression tree using geospatial analysis. J. Comput. Civ. Eng. **28**, 1–14 (2014). doi:10.1061/(ASCE)CP.1943-5487.0000432
4. Jadaan, K.S., Al-Fayyad, M., Gammoh, H.F.: Prediction of road traffic accidents in Jordan using Artificial Neural Network (ANN). J. Traffic Logistics Eng. **2**(2), 92–94 (2014)
5. Beynnon, B.D., Vacek, P.M., Newell, M.K., Tourville, T.W., Smith, H.C., Shultz, S.J., Johnson, R.J.: The effects of level of competition, sport, and sex on the incidence of first-time noncontact anterior cruciate ligament injury. Am. J. Sports Med. **42**(8), 1806–1812 (2014)
6. National Ski Areas Association: Skiergraphics: greater targeting for greater effectiveness. In: NSAA Convention and Tradeshow, Palm Springs, CA (2007)
7. Tuli, T., Haechl, O., Berger, N., Laimer, K., Jank, S., Kloss, F., Brandstätter, A., et al.: Facial trauma: how dangerous are skiing and snowboarding? J. Oral Maxillofac. Surg. **68**(2), 293–299 (2010)

8. Warme, W.J., Feagin Jr., J.A., King, P., et al.: Ski injury statistics 1982 to 1993 Jackson Hole Ski Resort. Am. J. Sports Med. **23**(5), 597–600 (1995)
9. Harlow, T.: Factors predisposing to skiing injuries in Britons. Injury **27**(10), 691–693 (1996)
10. King, M.A., Abrahams, A.S., Ragsdale, C.T.: Ensemble methods for advanced skier days prediction. Expert Syst. Appl. **41**(4), 1176–1188 (2014)
11. Demšar, J., et al.: Orange: data mining toolbox in Python. J. Mach. Learn. Res. **14**, 2349–2353 (2013)
12. Demšar, J., Zupan, B.: Orange: data mining fruitful and fun: a historical perspective. Informatica **37**, 55–60 (2013)
13. Quinlan, J.R.: C4.5 Programs for Machine Learning. Morgan-Kaufmann, San Francisco (1993)
14. Clark, P., Niblett, T.: The CN2 induction algorithm. Mach. Learn. **3**(4), 261–283 (1989)
15. Ishizaka, A., Nemery, P.: Multi-Criteria Decision Analysis: Methods and Software. Wiley, Chichester (2013)
16. Bohanec, M., Rajkovič, V., Bratko, I., Zupan, B., Žnidaršič, M.: DEX methodology: three decades of qualitative multi-attribute modelling. Informatica **37**, 49–54 (2013)
17. Figueira, J., Greco, S., Ehrogott, M.: Multiple Criteria Decision Analysis: State of the Art Surveys. Springer, New York (2005)
18. Costa, C.A.B.E., Vansnick, J.-C.: The MACBETH approach: basic ideas, software, and an application. In: Meskens, N., Roubens, M. (eds.) Advances in Decision Analysis, pp. 131–157. Kluwer Academic Publishers, Netherlands (1999)
19. Greco, S., Matarazzo, B., Słowiński, R.: Rough sets theory for multicriteria decision analysis. Eur. J. Oper. Res. **129**(1), 1–47 (2001)
20. Moshkovich, H.M., Mechitov, A.I.: Verbal decision analysis: foundations and trends. Adv. Decis. Sci. **2013**, 9 (2013)
21. Bohanec, M., Aprile, G., Costante, M., Foti, M., Trdin, N.: A hierarchical multi-attribute model for bank reputational risk assessment. In: Phillips-Wren, G., Carlsson, S., Respício, A., Brézillon, P. (eds.) DSS 2.0 – Supporting Decision Making with New Technologies, pp. 92–103. IOS Press, Amsterdam (2014)
22. Jakulin, A., Bratko, I.: Testing the significance of attribute interactions. In: Greiner R., Schuurmans D. (eds.) Proceedings of the Twenty-first International Conference on Machine Learning (ICML 2004), Banff, Canada, pp. 409–416 (2004)

The Effects of Performance Ratios in Predicting Corporate Bankruptcy: The Italian Case

Francesca di Donato and Luciano Nieddu[✉]

Unint - Università degli Studi Internazionali di Roma,
Via Cristoforo Colombo 200, 00145 Rome, Italy
{francesca.didonato,l.nieddu}@unint.eu

Abstract. Corporate failure prediction is a mayor issue in today's economy. Any prediction technique must be reliable (good recognition rate, sensitivity and specificity), robust and able to give predictions with a sufficient time lag to allow for corrective actions. In this paper we have considered the case of Small-Medium Enterprises (SMEs) in Italy trying to determine which dimension, in terms of performance indicators, best suits this goal. We have considered three of the most robust and diffused classification techniques on data over a period of 8 years prior to failure. The results tend to suggest that, for the Italian SME system, profitability ratios are always relevant in predicting corporate failure (both in the short and in the medium-long run), while leverage and liquidity indicators, affecting the financial dimension of the company, tend to add information in predicting a possible risk of default only in the medium-long run.

Keywords: Bankruptcy prediction · Discriminant analysis · Performance ratios · Small-medium enterprises

1 Introduction

An important body of accounting research has focused on the role of financial ratios for predictive purposes [1], in particular companies' failures. Since Beaver's [2] and Altman's [3] pioneering works, many studies have been devoted to exploring the use of accounting information in forecasting business failures. According to several authors (see, e.g., [3–5]), the failure of a limited company is related to two strictly connected situations: the inability to pay financial obligations when they come due (i.e. lack of liquidity and very high leverage) and the inability to generate operating profits (i.e. negative or very low income and profitability). The failure process is characterized by a systematic deterioration in the values of accounting ratios [6]. Sharma and Mahajan [7] presented a model of failure process according to which ineffective management together with unanticipated events lead to a systematic deterioration in performance ratios. In the absence of a corrective action this situation leads to failure.

While in the past, financial statement analysis was mainly used to assess the credit worthiness of borrowers, today, this analysis involves a wide variety of ratios and users. Firstly, banks and other financial institutions could avoid lending to failing firms. The financial investment sector could improve the risk return trade-off from investments by

© Springer International Publishing Switzerland 2015
B. Delibašić et al. (Eds.): ICDSST 2015, LNBIP 216, pp. 61–72, 2015.
DOI: 10.1007/978-3-319-18533-0_6

not investing in failing businesses. Moreover, companies could establish long-term relationships with other parties that will not likely fail in the future, and thus increase the longevity and viability of their business relationships. Finally, regulators could make early identifications of failing business in order to legally handle them, preventing illegal activities, such as avoiding taxes or diluting debt [8].

One relevant issue is the existence of many definition of "failure" in literature. Many studies define failure as actual filing for bankruptcy or liquidation (see, e.g. [3, 9, 10]). Others define it as suffering financial stress or an inability to pay financial obligations [2, 11]. Some authors use the term 'failed' interchangeably with 'bankrupted' (see [12]). When financial distress cannot be relieved, it will lead to bankruptcy even if it is difficult to discern the precise moment that bankruptcy occurs. McKee [13] stated that a firm goes through various stages of financial distress before bankruptcy (i.e. inadequate income and liquid asset position, difficulties with paying the invoices).

To predict bankruptcy, Beaver [2] used only univariate statistics on US market data, finding out a high predictive ability of financial ratios up to 5 years before failure. The capability of univariate analysis to assess potential bankruptcy of firms is questionable since it focuses on individual signals of impending problems [3]. For this reason, Altman [3], applied Linear Discriminant Analysis (LDA) [14] and his approach was popularised as the Z-score model. In his model no cash flow ratios were found to be significant, contrasting with Beaver's analysis. Altman's LDA outperformed Beaver's model for one year prediction intervals. He went back up to 5 years prior to failure but the results deteriorated already at 3 years yielding a prediction rate below 50 %. Upon Altman's model, many studies have been conducted in the USA examining the usefulness of models based on accounting data for the prediction of big corporate failure [4, 9, 15], in different industries as railroads [16], banks [17] and insurance sector [18, 19].

The interest in this approach has spread outside the USA in the late 70s. Taffler [20–23] developed a UK-based Z-score model to analyse the financial health of firms listed on the London Stock Exchange. Other studies were drawn for French [24] and Australian companies [25]. Before Storey's [26] work, only few studies dealt with the failure of small-medium firms (SMEs) [27, 28]. Storey [26] drew his sample from the SME sector, also using non financial variables but without considering a control group of surviving firms. Hall [29] studied the factors affecting small companies failure distinguishing failing and surviving firms but only considering the construction sector. Hence, to our knowledge, there are few studies in the literature focusing on SMEs. For this kind of firms it would be very helpful to have some tools to predict their possible failure in advance and use this information to make management decisions. SMEs are often under-capitalized and mostly rely on external financial sources provided by banks and other financial institutions, which then need prediction models to evaluate the risk of a loan loss.

It is also important to consider how far ahead the model is able to accurately predict bankruptcy. Many studies have a high accuracy rate one year prior to failure, however some models are able to predict bankruptcy much sooner. For example, Deakin's model [15] could predict bankruptcy with 96 % accuracy 2 years prior to failure, even if such short time frame was judged, by an expert banking panel, not to be sufficiently timely, in general, to allow a lending institution to extricate itself without incurring the risk of a significant loan loss [30]. In general, the farther back in time we go the less the accuracy level of the model is and therefore its usefulness. Blum [9] and Laitinen [6], predicted bankruptcy 6 years prior to failure but with only 57 % and 60 % accuracy. The main goal of our paper is to investigate if it is possible to develop reliable failure prediction models for the Italian public industrial SMEs, and to explore the incremental information content of accounting ratios in predicting financial collapse over time. On a second stage we want to determine which is the best prediction model for the Italian case at various stages in time prior to failure among those mostly used in the literature.

The results concern a retrospective of the financial statements of 50 active firms and 50 failed firms randomly selected over a period of 12 years, computing a set of ratios selected by popularity. Apart from the study of Appetiti [31], no further attempts have been made to develop failure models in Italy for SMEs representing around 90 % of the overall Italian enterprises. In Italy these companies mainly depend on external source of finance, basically provided by banks. In particular, due to the current financial crisis, the risk was that, in such a bank-oriented industry, any subsequent contagion on the inter-banking market side would be able to jeopardise the principal source of external finance for the firms, because of banks tightening the credit access to their borrowers. In such a context, the "size effect" plays a relevant role. In fact there is a large amount of theoretical and empirical literature focusing on the SMEs gap because these firms have more diffi-culties in getting external finance, mainly due to problems of information asymmetries.

It must be stressed that in the literature there is a number of approaches to bankruptcy prediction assessment. Although it was pointed out by Jones [32] that any methodology to predict bankruptcy should be tested on a hold-out sample of firms not used in the analysis, this suggestion has not been utterly followed. Any methodology tested on the dataset that it has been trained on, yields biased performance estimates (see e.g. [33] or [34]). Even when the methods have been tested on a hold-out sample the result is dependent on the particular sample that has been held-out. In the following experimental setup the methodologies will be tested using leave-one-out (LOO) [35], i.e. all the firms in turn will be tested.

The layout of the paper is as follows. In Sect. 2 the materials and methods used in this paper will be described. Section 3 will contain the results and in Sect. 4 the discussion will be presented. Finally in Sect. 5 some conclusions will be drawn.

2 Materials and Methods

In this study, we employ an original data set consisting of 100 non-listed SMEs during the years 2000–2011, 50 that filed for bankruptcy and 50 still operating at the end of 2011, using business sector as stratifying variable, choosing only firms with turnover in the range 2–50 million euros at the beginning of the analysed period. The sample was randomly selected from the firms operating in Italy. The reason for the analysis is to determine if the selected ratios entail different information content with respect to their ability to predict bankruptcy. As in Abdullah et al. [36], financial companies and property industries were not considered in the analysis since their ratios are highly volatile. Besides, the interpretation of the ratios is slightly different since financial companies, for example, have different nature of income and expenses from non-financial companies.

Due to the longitudinal nature of the data, the sample dimension will decrease from 100 companies at 2000 to a plateau of 50 companies at 2011. The aim of the study is to describe which dimensions can be considered as indicators of possible distress for a firm.

Two questions emerge when attempting to select accounting ratios for empirical research: 1. Which ratios, among the literally hundreds available, should be used? 2. Which prediction model is best fit for the problem at hand? The theoretical models provide little foundation as a guide in the choice.

Altman and Saunders [37] and Allen et al. [38] reviewed the vast literature on the influence of accounting indicators on corporate distress in detail. They identify the predominant use of LDA and logistic models in corporate distress prediction and the influence of several ratios. The data used in this work were collected through CERVED database, related to economic and financial data of Italian non-listed companies. Using the available financial statements of each firm, the most common ratios for every year in the period 2000–2011 have been computed. The lack of any real theory related to the use of accounting ratio analysis constitutes a serious gap in the accounting literature, apart from the paper of Lev and Sunders [39] and Whittington [40]. According to Barnes [1] we selected the ratios throughout the criterion of popularity, meaning their frequency of appearance in the literature [41]. We have grouped the selected ratios according to the economic or financial dimension of the company into three groups:

1. profitability ratios alone, related to the economic dimension of the company: Return on Equity (ROE), Capital Turnover (CT), Net Income/Total Assets (NI_TA), Return on Investment (ROI), Earning/Sales (ES), Return on Sales (ROS), Financial Interests/Ebitda (FI_EB), Financial Interest/Sales (FI_S);
2. leverage and liquidity ratios alone, related to the financial dimension of the company: Financial Debts/Equity (FDE), Short Term Bank Loan/Working Capital (STBL_WC), Cash Flow/Total Debt (CF_TD), Structure Ratio 1 (ST1), Structure Ratio 2 (ST2), Working Capital/Total Assets (WC_TA), Quick Ratio (QR), Working Capital Cycle (WCC), Financial Debt/Working Capital (FD_WC), Current Ratio (CR), Retained Earnings/Total Assets (RE_TA);
3. overall performance indicators (Overalls), considering profitability, liquidity and leverage ratios at the same time (economic and financial dimension).

To evaluate the predictive ability of the performance ratios, three among the most used classification techniques in the sector have been applied, namely LDA, Quadratic Discriminant Analysis (QDA) and Classification Trees (CARTs) (see e.g. [33, 34]). Since 1966 until recently LDA was the dominant method in failure prediction. As already mentioned, Altman [3] was the first to apply LDA to predict the failures of firms from different industries. To make the paper self-contained, a brief overview of the theory of these classification algorithms follows.

In a standard classification problem, the outcome of interest falls into J unordered and mutually exclusive classes, which can be denoted by the set $\{1,2,3, ..., J\}$. The aim of any classification algorithm is then to build a rule $h(.)$, based on the data at hand (training set), to predict the class membership of an item based on p measurements of predictors (features) $x \in \mathbb{R}^p$. In the specific case the aim is to classify a firm into one of two classes, (healthy and distressed) based on a set of performance indicators.

The problem that is posed in LDA is to determine which linear combination $\mathbf{wx}$ (projection) of the p variables best separates the sample of failed firms from that of healthy ones. The results are evaluated with respect to a criterion, usually involving some ratio of the within-groups, between-groups variances. The solution outlined by Fisher (1936) consists in determining a projection such that the ratio of the difference between sample means to the within-groups standard deviation is maximized, i.e.:

$$\text{argmax}_w \frac{\mathbf{w}^t S_b \mathbf{w}}{\mathbf{w}^t S_w \mathbf{w}}. \tag{1}$$

The solutions to this problem can be found in the eigenvectors of the matrix $S_w^{-1} S_b$ (at most $J - 1$ non zero eigenvectors). The original approach due to Fisher can be considered a dimensionality reduction method that preserves as much of the class separation as possible.

In a parametric framework, if the data have been drawn from multivariate normal populations with equal covariance matrices, the solution to the maximization problem (1) also provides the best bayesian classifier (see [33, 34]) where units are assigned to the class k that maximizes the posterior probability, i.e.:

$$\text{argmax}_w \left\{ -\frac{1}{2} \ln |\Sigma_k| - \frac{1}{2} (x - \mu_k)' \Sigma_k^{-1} (x - \mu_k) + \ln \pi_k \right\}$$

Where Σ_k is the covariance matrix for class k and π_k is the prior probability of vector x to belong to class k. If the data are drawn from homoscedastic populations (equal covariance matrices Σ_k for all classes) this rule reduces to a linear discriminant rule which is equivalent to the LDA classifier. When the homoscedasticity assumption is violated [42] a quadratic classifier can be estimated (QDA).

Since LDA can be considered a special case of QDA, it goes without saying that the latter allows for more flexibility for the covariance matrix and therefore tends to better fit the data. This increased flexibility is paid in term of a larger number of parameters to estimate since a separate covariance matrix for each class must be estimated from the data. This could be a problem when the number of elements in the training set is not large.

An alternative approach to LDA and QDA, in presence of missing data or when the number of features is too large when compared to the number of cases in the dataset, is represented by Classification Trees. Classification and Regression Trees (CARTs) have been popularized by Breiman et al. [43] and are a particular case of cluster weighted modeling [44] where the value of an outcome variable must be predicted according to the values assumed by a set of predictors. The basic idea for CARTs is to partition the feature space in non-overlapping regions and then fit a separate model in each region for the outcome variable. The problem of detecting the best partition on the feature space that optimizes a measure of impurity on the subsets is NP-hard. Various heuristics exit to get a local optimal solution to the problem.

Once a classifier has been trained on the available dataset, its performance must be determined. The performances of the various classifiers will be evaluated according to correct recognition rate, sensitivity and specificity. Namely, consider a statistical test that allows choosing between two hypotheses (H_0 and H_1). Let H_0 be "the firm belongs to the non-failed ones" and H_1 be "the firm belongs to the failed ones":

- *sensitivity* of a test is the statistical power of the test and is related to the type II error (non rejecting H_0 when it is false). It can be estimated using the proportion of firms that have failed that are actually recognized by the test. A highly sensible test is good for ruling out the condition under testing. Positive results in a highly sensitive test are not useful to rule in the firm as being failed since sensitivity does not take into account false-positives. A test that would classify all firms as having failed would have a perfect sensitivity but would not be useful in determining those that actually fail.
- *specificity* of a test is related to the type I error of a statistical test (rejecting the null hypothesis when it is true). It can be estimated using the proportion of sound firms that have been recognized as "healthy" by the test. Specificity is therefore the ability of a test to exclude a condition correctly. Specificity is not useful for ruling out a hypothesis. A fake test that would classify all firms as healthy would have a perfect specificity. If a firm test positive (failed) to a highly specific test than it would have a great probability of being a failed firm.
- *correct recognition rate* of a test is the probability of correctly classifying a new element. It can be estimated using the proportion of correctly classified firms over the total number of firms.

Unbiased estimates of these quantities can be obtained via cross-validation, i.e. part of the sample is selected to train the classifier and part, independent of the previous one, is used to assess the performance of the classifier. Usually a k-fold cross-validation scheme is used. A special version of the k-fold cross validation is the LOO scheme, where in turn each element of the sample is singled out to be tested on the classifier trained on the remaining n-1 elements. The performance of the classifier is then a synthesis of the outcomes on each unit. LOO is particularly useful when the dataset is not large and therefore as much as possible elements of the dataset should be used in training.

Although the data is longitudinal in nature, the study we have carried out is a cross-sectional study: the failed companies have been considered at various years prior to failure. Each distressed company was randomly matched with a healthy company belonging to the same industry sector and had the closest total assets. The criteria were set as a control factors to ensure minimum bias in the selection of the control sample used in the development of the failure prediction model. Due to the casuality of the matching mechanism, the selection mechanism has been repeated 300 times to get an average estimate of the performance of each prediction technique. Companies have been considered from 1 up to 8 years prior to failure. Similar studies have only considered data up to 6 years prior to failure (see, e.g. [41] for a detailed review).

3   Results

In Table 1 the average correct recognition rates (RR), sensitivities (Sens) and specificities (Spec) over 300 trials for the three considered methods have been displayed.

As it was to be expected, the average correct recognition rate decreases as the time lag increases. No results have been obtained for QDA with a time lag of 8 years on the overall set of indicators, since the sample size is too small and the number of variables is to large to get estimates for all the parameters in the model (see Sect. 2).

Table 1. Average recognition rates (RR) Sensitivities (Sens) and Specificities (Spec) over 300 replications.

Type	Time Lag (years)	CARTs			LDA			QDA		
		RR	Sens	Spec	RR	Sens	Spec	RR	Sens	Spec
Overall	1	**0.916**	0.930	0.902	0.765	0.641	0.888	0.870	**0.995**	0.824
	2	0.849	0.834	0.863	0.767	0.665	0.869	0.845	**0.977**	0.857
	3	0.811	0.768	0.854	0.833	0.718	**0.948**	0.842	**0.977**	0.859
	4	0.810	0.802	0.818	0.792	0.746	0.838	0.867	**0.974**	0.872
	5	0.793	0.739	0.846	**0.844**	0.805	0.883	0.835	**0.942**	0.842
	6	0.712	0.663	0.762	**0.818**	0.749	**0.887**	0.777	**0.960**	0.657
	7	0.734	0.699	0.769	**0.763**	0.712	**0.815**	0.445	0.826	0.410
	8	0.630	0.588	0.672	**0.660**	0.569	**0.750**	--	--	--
Financial	1	0.816	0.850	0.781	0.677	0.748	0.606	0.619	0.311	**0.940**
	2	0.752	0.771	0.733	0.742	0.725	0.758	0.614	0.302	**0.933**
	3	0.738	0.781	0.695	0.762	0.801	0.723	0.554	0.209	0.908
	4	0.734	0.728	0.739	0.621	0.638	0.604	0.533	0.110	**0.954**
	5	0.684	0.638	0.730	0.653	0.799	0.508	0.634	0.566	0.715
	6	0.662	0.640	0.683	0.623	0.727	0.519	0.639	0.686	0.593
	7	0.686	0.685	0.687	0.745	**0.861**	0.629	0.701	0.768	0.640
	8	0.635	0.624	0.647	0.654	0.739	0.569	0.553	0.623	0.487
Profitability	1	**0.916**	0.930	0.902	0.735	0.575	0.894	0.896	0.955	0.913
	2	0.850	0.837	0.863	0.753	0.630	0.876	**0.882**	0.912	0.925
	3	0.820	0.779	0.861	0.806	0.703	0.908	**0.882**	0.909	0.897
	4	0.818	0.815	0.821	0.814	0.780	0.848	**0.877**	0.928	0.870
	5	0.797	0.750	0.844	0.814	0.777	0.850	0.813	0.779	**0.888**
	6	0.723	0.660	0.786	0.799	0.783	0.814	0.792	0.870	0.750
	7	0.721	0.693	0.748	0.714	0.704	0.724	0.724	0.820	0.717
	8	0.624	0.576	0.673	0.655	0.611	0.699	0.626	**0.872**	0.522

In bold the best performances for each combination of time-lag, set of indicators and classification algorithm have been displayed. From the results it seems that CARTs represent the best prediction algorithm in the immediate proximity of a failure, with an average recognition rate of 0.916 one year prior to failure, both on profitability ratios and on the overall set of ratios. CARTs have the overall best performance one or two years prior to failure on the three sets of ratios. Considering a time lag from 2 to 4 years prior to failure, the best performance, in term of recognition rates, is obtained using QDA on only profitability indicators. Therefore, on a short time horizon, up to 4 years prior to failure, the profitability ratios are those that provide the best performance. In the medium-long run (5–8 years prior to failure) the best performance is obtained using LDA on the total set of indicators (overall). Since the profitability indicators are a subset of the overall set of indicators, their improved performance in the short run, when compared with the overalls, could be due to a noise effect of the liquidity and leverage ratios that could have no further informative power in determining a stressful situation for the firm in such a time horizon. In the medium-long run the overall set of indicators shows a better performance than either the profitability or liquidity-leverage ratios when taken separately. Therefore these last indicators add information, for the survival of the firm, to the one already contained in the profitability indicators only in the medium-long run.

Table 1 details also the average sensitivities for the same set of experiments. Sensitivities are the probabilities of predicting a firm as failed given that the firm has failed. It is also known as test power. The result of a highly sensitive test can be used as an indicator of the health of the firm, since such a test can be used to rule out the condition under testing (failure). From the obtained results, the most sensitive classifiers (i.e. the most powerful) are those that use QDA on the overall set of indicators up to 6 years prior to failure. For 7 and 8 years the most powerful are LDA on financial indicators and QDA on economic indicators respectively. The behavior of CARTs is somehow erratic. This was to be expected since CART is a non parametric technique and therefore less powerful then any parametric classifiers.

Finally, the average specificities and their standard deviations over the 300 trials have been displayed. A test with high specificity is good as a warning signal, since a firm that is classified as failed by a highly specific test is likely to actually be a failed firm. The inverse is not true, i.e. a company that is classified as healthy by a highly specific test does not mean that it will not fail. The most specific classifiers are those using QDA on indicators affecting the financial dimension of the company in the short run and LDA on the overall set in the medium-long run. As it was to be expected, a highly specific classification algorithm in the long run must take into account both profitability and liquidity-leverage since the health of a firm is measured by those two dimensions at the same time. Moreover, this is consistent with the fact that the lack of cash in the short term (financial dimension) comes from shortage of revenues affecting at the same time the economic and financial dimension of the company in the medium-long term.

The higher the level of revenue financing, the less the firm needs outside funds in the form of debt and share capital, lowering its financial obligations. Only one of these two dimensions can be of importance in a short time window, but both aspects must be considered to obtain a complete measure of the health of the firm.

4   Discussion

This discussion considers the methodological aspects of the analysis of failure data and the results we have obtained in the experimental phase. In the last decades many different approaches to failure prediction have been developed [41]. There is no general agreement on how to test each single approach. The use of holdout sample and cross-validation is not so frequent in the literature as it should be. Although in the specialized literature it was suggested the need of an independent sample to test the classifier [32], several works have continued testing the performance of various techniques on different sets of variables using only resubstitution error. The estimates obtained using resubstitution are biased estimates of the real performance, giving an error which is, on average, lower than the actual error. Bellovary et al. [41] report that from 1987 to 2007, out of 90 papers on the topic, almost 45 % still did not use hold-out sample or cross-validation to test the performances. Any algorithm trained on a dataset will perform well on the dataset it has been trained on. To get a glimpse at the actual potential performance of the proposed methodology a cross-validation approach must be used which provides a nearly unbiased estimate [45] of the future error rate. With LOO this small bias is further reduced. Besides, using LOO or k-fold cross-validation [33] assures that all the elements of the dataset will in turn be tested, avoiding any subjectivity in the choice of the hold-out sample.

In this paper the performance of the various methods on the different sets of indicators is consistent with the literature and depicts a picture of the Italian SMEs which is somehow peculiar. In fact, although the best warning signal for bankruptcy is given by QDA on financial indicators (Table 1) for the short time, it has very low sensitivity, at risk of non-detecting firms that are likely to fail. The profitability ratios on QDA in the short period have both quite high sensitivity and specificity and an optimal error rate. For the Italian case, they have the highest prediction power and are good at providing a nice trade off between the ability to detect firms that are likely to fail in the short time and to rule out firms that will not. The importance of these ratios is consistent with Altman [3] and Pinches and Trieschmann [18] because the profitability of bankrupted firms is almost nil. Revenues are one of the prime sources of funds used to repay debts. The funds from operations will be equal to the total revenues earned during the period minus the expenses incurred for the period [46]. It goes without saying that a firm can be able to operate in the short term without considering the financial dimension of the operations but in stressful situation and in the medium-long term, the access to credit and financial aid is mandatory to be able to keep operating.

This is consistent with the results showing that in the medium-long run the most powerful method (the one with the highest sensitivity) is QDA on the overalls (profitability, liquidity and leverage) considering both dimensions of the company at the same time. This has been also stressed out by Taffler [23] stating that the evaluation of firms' performance only considering one single dimension, such as its profitability or its liquidity alone, overlooking the other dimensions, can lead to an incomplete and potentially misleading view. The different ratios are each measuring a distinct aspect of a firm. Therefore, in the medium–long term liquidity and leverage ratios add information to the survival of the firm, i.e. in the medium-long term, survival is explained by the overall ratios.

These results can be explained by the characteristics of the Italian SMEs which are very often under capitalized and have a financial structure mainly based on financial liabilities. For these reasons leverage and liquidity ratios tend to be not very good mainly for a physiological condition due to the company's financial structure. That is why leverage and liquidity ratios alone could not be considered as efficient predictors of companies' failure since their values are strictly related to a typical characteristic of Italian SMEs. On the other side, they could add information – in the medium-long run – when considered with profitability ratios, which are, anyhow, the most important ratios in predicting corporate failure. This is consistent with Bierman [46], because the liquidity of a firm depends mainly on its profitability, that is the sufficiency of revenues financing stemming from sales. The higher the level of revenue financing, the less the firm needs outside funds. Therefore profitability ratios always matter in the short and in the medium-long period.

5 Conclusions

We performed a cross sectional study based on a sample of 100 Italian non listed SMEs over the period from 2000 to 2011, considering 50 firms that have declared bankruptcy during this time period and 50 still active on the market at 2011. The focus was to determine which dimensions, in term of sets of performance indicators, best predict the possible failure of a firm in Italy. Three techniques have been compared to find out which one yields the best performance in terms of correct recognition rate, sensitivity and specificity on three sets of performance ratios.

The best trade off between correct recognition rate, sensitivity and specificity is obtained on the LDA on the overall set of indicators in the long run. The recognition rates are quite good, and the predictive ability of the methods has been tested up to 8 years prior to failure. These results are in line with those available in the literature going from 90 % correct recognition rate one year prior to failure to 66 % 8 years prior to failure. As a peculiar feature of the Italian SME system, the data show an informative power of the accounting indicators as a warning signal to predict bankruptcy only the medium-long run.

References

1. Barnes, P.: The analysis and use of financial ratios: a review arude. J. Bus. Finance Acc. **14**(4), 449–461 (1987). Winter
2. Beaver, W.H.: Financial ratios as predictors of failure. JAR **4**, 71–111 (1966)
3. Altman, E.I.: Financial ratios, discriminant analysis and prediction of corporate bankruptcy. J. Finance **23**(4), 589–609 (1968)
4. Ohlson, J.A.: Financial ratios and probabilistic prediction of bankrucy. JAR **18**(1), 109–131 (1980)
5. Bhimani, A., Gulamhussen, M.A., Da Rocha Lopes, s: The role of financial, macroeconomic, and non-financial information in bank loan default timing prediction. Eur. Acc. Rev. **22**(4), 739–763 (2013)

6. Laitinen, E.K.: Financial ratios and different failure processes. J. Bus. Finance Acc. **18**(5), 649–673 (1991)
7. Sharma, S., Mahajan, V.: Early warning indicators of business failure. J. Mark. **44**(4), 80–89 (1980)
8. Gepp, A., Kumar, K.: Business Failure prediction using statistical techniques: A review, Bond University ePublication (2012)
9. Blum, M.: Failing company discriminant analysis. JAR. **12**, 1–25 (1974)
10. Boardman, C.M., Bartley, J.W., Ratliff, R.B.: Small business growth characteristics. Am. J. Small Bus. **5**, 33–42 (1981)
11. Chen, K.C., Lee, C.J.: Financial ratios and corporate endurance: A case of the oil and gas industry. Contemp. Acc. Res. **9**(2), 667–694 (1993)
12. Karels, G.V., Prakash, A.J.: Multivariate normality and forecasting of business bankruptcy. J. Bus. Finance Acc. **14**(4), 573–593 (1987)
13. McKee, T.E.: Rough sets bankruptcy prediction models versus auditor signaling rates. J. Forecast. **22**, 569–586 (2003)
14. Fisher, R.A.: The use of multiple measurements in taxonomic problems. Ann. Eugen. **7**, 179–188 (1936)
15. Deakin, E.B.: A discriminant analysis of predictors of business failure. JAR **10**(1), 167–179 (1972)
16. Altman, E.I.: Predicting railroad bankruptcies in America. Bell J. Econ. Manage. Sci. **4**(1), 184–211 (1973)
17. Sinkey Jr., J.: A multivariate statistical analysis of the characteristics of problem banks. J. Finance **30**(1), 21–36 (1975)
18. Trieschmann, J.S., Pinches, G.E.: A multivariate model for predicting financially distressed P-L insurers. J. Risk Insur. **40**, 327–347 (1973)
19. Pinches, G.E., Trieschmann, J.S.: Discriminant analysis, classification results and financial distressed P-L insurers. J. Risk Insur. **44**, 289–298 (1977)
20. Taffler, R.J.: Finding those companies in danger using discriminant analysis and financial ratio data: a comparative based study city business school. Working Paper No. 3, City University Business School, London (1977)
21. Taffler, R.J., Tisshaw, H.: Going, going, gone – four factors which predict. Accountancy, March, 50–54 (1977)
22. Taffler, R.J.: Forecasting company failures in the UK using discriminant analysis and financial ratio data. J. R. Stat. Soc. Ser. A **145**(3), 342–358 (1982)
23. Taffler, R.J.: The assessment of company solvency and performance using a statistical model: a comparative UK based study. Acc. Bus. Res. **13**(52), 295–307 (1983)
24. Altman, E.I., Margaine, M., Schlosser, M., Vernimmen, P.: Financial and statistical analysis for commercial loan evaluation. JFQA **9**(2), 195–211 (1974)
25. Castagna, A.D., Matolcsy, Z.P.: The prediction of corporate failure: testing the Australian experience. Aust. J. Manage. **6**(1), 23–50 (1981)
26. Storey, D., Keasey, K., Watson, R., Wynarczk, P.: The Performance of Small Firms. Croom Helm, London (1987)
27. Edmister, R.: An empirical test of financial ratio analysis for small business failure prediction. J. Finance Quant. Anal. **7**(2), 1477–1493 (1972)
28. Argenti, J.: Corporate Collapse. The Causes and Symptom. McGraw-Hill, New York (1976)
29. Hall, G.: Factors distinguishing survivors from failures amongst small firms in the UK construction sector. J. Manage. Stud. **31**(5), 737–760 (1994)
30. Casey, C.J.: The usefulness of accounting ratios for subjects' predictions of corporate failure: replication and extensions. JAR. **18**(2), 603–613 (1980)

31. Appetiti, A.: Identifying unsound firms in Italy: an attempt to use trend variables. J. Bank. Finance **8**(2), 269–279 (1984)
32. Jones, F.L.: Current techniques in bankruptcy prediction. J. Acc. Lit. **6**, 131–164 (1987)
33. Nieddu, L., Patrizi, G.: Formal methods in pattern recognition: a review. EJOR. **120**(3), 459–495 (2000)
34. McLachlan, G.J.: Discriminant Analysis and Statistical Pattern Recognition. Wiley, New York (2004)
35. Efron, B., Tibshirani, R.: Cross-validation and the bootstrap: estimating the error rate of a prediction rule. Technical report 176, Stanford University (1995)
36. Abdullah, N.A.H., Halim, A., Rus, R.M.D.: Predicting corporate failure of Malaysia's listed companies: comparing multiple discriminant analysis, logistic regression and the hazard model. International Research Journal of Finance and Economics **1**(15), 201–217 (2008)
37. Altman, E.I., Saunders, A.: Credit risk measurement: developments over the last 20 years. J. Bank. Finance **20**(1), 1721–1742 (1998)
38. Allen, A., Delong, G., Saunders, A.: Issues in the credit risk modeling of retail markets. J. Bank. Finance **28**(1), 727–752 (2004)
39. Lev, B., Sunder, S.: Methodological issues in the use of financial ratios. J. Acc. Econ. **1**, 187–210 (1979)
40. Whittington, G.: Some basic properties of accounting ratios. J. Bus. Finance Acc. **7**(2), 219–232 (1980)
41. Bellovary, J.L., Giacomino, D.E., Akers, M.D.: A review of bankruptcy prediction studies: 1930 to present. J. Financ. Educ. **33**, 1–42 (2007). winter
42. Smith, C.A.B.: Some examples of discrimination. Ann. Eugen. **13**, 272–282 (1947)
43. Breiman, L., Friedman, J.H., Olshen, R.A., Stone, C.J.: Classification and Regression Trees. Wadsworth, Belmont (1984)
44. Gershenfeld, N.: Nonlinear inference and cluster-weighted modeling. Ann. N. Y. Acad. Sci. **808**, 18–24 (1997)
45. Hastie, T., Tibshirani, R., Friedman, J.: The Elements of Statistical Learning. Data Mining, Inference, and Prediction. Springer, New York (2001)
46. Bierman Jr., H.: Measuring financial liquidity. Account. Rev. **35**(4), 628–632 (1960)

A Tangible Collaborative Decision Support System for Various Variants of the Vehicle Routing Problem

Nikolaos Ploskas[1]([envelope]), Ioannis Athanasiadis[2], Jason Papathanasiou[1], and Nikolaos Samaras[1]

[1] University of Macedonia, 156 Egnatia Street, 54006 Thessaloniki, Greece
{ploskas,jasonp,samaras}@uom.gr
[2] Hellenic Open University, 8 Ptolemaion Street, 50100 Kozani, Greece
g-athan@hotmail.com

Abstract. The Vehicle Routing Problem (VRP) is a well-known combinatorial optimization problem where a number of customers must be served with a fleet of vehicles. The classical variation of the VRP is the Capacicated Vehicle Routing Problem (CVRP) with the additional constraint that each vehicle must have uniform capacity. Many Decision Support Systems (DSS) have been implemented to solve real life problems of the VRP and its' variants, but they do not allow multiple decision makers to explore several scenarios of a given problem simultaneously and collaborate with each other in order to find the best possible solution. In this paper, we extend our previous work [35] and incorporate in our spatial DSS four variants of the CVRP: (i) Distance Constrained Vehicle Routing Problem (DVRP), (ii) Vehicle Routing Problem with Time Windows (VRPTW), (iii) Vehicle Routing Problem with Backhauls (VRPB), and (iv) Vehicle Routing Problem With Pickup and Delivery (VRPPD). This extension allows decision makers to solve specific routing problems according to their needs. The proposed collaborative spatial DSS allows two decision makers to collaborate with each other in order to find the best possible solution through a tangible interface. The locations are added through interactive Google Maps and other parameters through user-friendly forms that can be manipulated via tangible interfaces. The proposed DSS has been implemented using Java, TUIO protocol, jsprit, and Google Maps.

Keywords: Decision Support Systems · Capacitated Vehicle Routing Problem · Tangible user interface · Geographical information systems · Tabletop display

1 Introduction

The Vehicle Routing Problem (VRP) is a well-known combinatorial optimization problem in the fields of transportation, distribution and logistics [24,46]. VRP has been initially introduced by Dantzig and Ramser [10]. In the classical VRP,

© Springer International Publishing Switzerland 2015
B. Delibašić et al. (Eds.): ICDSST 2015, LNBIP 216, pp. 73–84, 2015.
DOI: 10.1007/978-3-319-18533-0_7

a fleet of vehicles, which are located at a central depot, must deliver a given quantity of products to a number of customers. The objective is to determine the optimal route for a number of vehicles that will serve all customers (and each customer will be served only once) by minimizing the overall transportation cost. The majority of the real world problems are more complex than the VRP. So, many variants of the VRP have been proposed. The Capacitated Vehicle Routing Problem (CVRP) is the most classical variant of the VRP adding an additional constraint that each vehicle must have uniform capacity. As shown in Fig. 1, the most well-known variants of the CVRP are: (i) Distance Constrained Vehicle Routing Problem (DVRP), (ii) Vehicle Routing Problem with Time Windows (VRPTW), (iii) Vehicle Routing Problem with Backhauls (VRPB), and (iv) Vehicle Routing Problem With Pickup and Delivery (VRPPD).

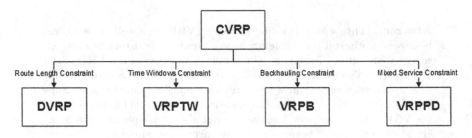

Fig. 1. CVRP variants

Many Decision Support Systems (DSS) exist for the solution of real life problems of the VRP and its' variants. Only few of them integrate real-life geographical information [2,18,19,37,39,44]. Anderson et al. [1] used tangible interfaces for the solution of the CVRPTW through a tabletop display. In our previous paper [35], we proposed a collaborative spatial DSS for the CVRP on a tabletop display. In this paper, we extend our previous paper [35] and incorporate in our spatial DSS four variants of the CVRP: (i) DVRP, (ii) VRPTW, (iii) VRPB, and (iv) VRPPD. Our previous experience [35] showed that decision makers need to solve specific routing problems, so the addition of these variants will offer them the opportunity to solve their routing problems by representing the real market and not making any abstractions. The proposed collaborative spatial DSS allows two decision makers to collaborate with each other in order to find the best possible solution through a tabletop display.

The structure of the paper is as follows. Section 2 briefly presents the mathematical form of the CVRP and its' four aforementioned variants. Section 3 presents some key features about the tangible user interfaces, while in Sect. 4 the analysis and implementations steps of the collaborative spatial DSS are presented. Finally, the conclusions of this paper are outlined in Sect. 5.

2 Problem Specification

The classical variation of the VRP is the CVRP. The CVRP is the generalization of the Traveling Salesman Problem (TSP), where the products are to be delivered to a number of customers by a fleet of identical vehicles starting and ending at a central depot. The objective is to determine a viable route schedule, which minimizes the distance or the total cost with the following constraints:

- Each vehicle starts and ends its route at the central depot.
- Each customer should be served once by one vehicle.
- The total demand of each route must not exceed the capacity of each vehicle.

Let us assume that the central depot is node 0 and N customers are served by V vehicles. The demand of customer i is d_i and the capacity of vehicle v is c_v. The cost traveling from customer i to customer j by vehicle v is C_{ij}^v. The mathematical form of this problem based on the formulation given by Bodin et al. [5] can be formulated as follows [35]:

$$min \sum_{v=1}^{V} \sum_{i=0}^{N} \sum_{j=0}^{N} C_{ij}^v X_{ij}^v \tag{1}$$

subject to

$$X_{ij}^v = \begin{cases} 1, \text{ if vehicle v travels from customer i to j} \\ 0, \text{ otherwise} \end{cases} \tag{2}$$

$$\sum_{v=1}^{V} \sum_{i=0}^{N} X_{ij}^v = 1, j = 1, 2, ..., N \tag{3}$$

$$\sum_{v=1}^{V} \sum_{j=0}^{N} X_{ij}^v = 1, i = 1, 2, ..., N \tag{4}$$

$$\sum_{i=0}^{N} X_{it}^v - \sum_{j=0}^{N} X_{tj}^v = 0, v = 1, 2, ..., V \text{ and } t = 1, 2, ..., N \tag{5}$$

$$\sum_{j=0}^{N} c_j \left(\sum_{i=0}^{N} X_{ij}^v \right) \leq c_v, v = 1, 2, ..., V \tag{6}$$

$$\sum_{i=1}^{N} X_{0j}^v \leq 1, v = 1, 2, ..., V \tag{7}$$

$$\sum_{j=1}^{N} X_{i0}^v \leq 1, v = 1, 2, ..., V \tag{8}$$

Objective function (1) refers to the minimization of the total cost. Constraint (2) ensures that the variable X_{ij}^v takes the integer 0 or 1. Constraints (3) and (4) ensure that each customer is served once, while Constraint (5) ensures the route continuity. Constraint (6) ensures that the total demand of each route will not exceed the capacity of each vehicle, while Constraints (7) and (8) ensure that each vehicle is used only once.

Exact and heuristic algorithms exist for the solution of the CVRP. In the first category, branch-and-cut methods [4,29] and the branch and-cut-and-price algorithm proposed by Fukasawa et al. [14] have been proposed (for a detailed survey of exact algorithms for the CVRP, see [32,47]). In the second category, many heuristics and metaheuristics have been proposed (for a detailed survey of approximate algorithms for the CVRP, see [7,27]).

The following sub-sections of this section present the four variants of the CVRP that are incorporated in our proposed DSS: (i) DVRP, (ii) VRPTW, (iii) VRPB, and (iv) VRPPD.

2.1   Distance Constrained Vehicle Routing Problem

The DCVRP is a variant of the CVRP adding the following constraint: the total length of each route must not exceed a fixed length. Based on the previous formulation, let us denote with L_v the maximum allowed total length of the route served by vehicle v. Then, one more constraint, which refers to the maximum allowed fixed length of each route, should be added to the formulation of the CVRP in order to represent the DCVRP variant:

$$\sum_{i=0}^{N} \sum_{j=0}^{N} d_{ij}^v X_{ij}^v \leq L_v, \text{v} = 1, 2, ..., \text{V} \tag{9}$$

Exact and heuristic algorithms exist for the solution of the DCVRP. In the first category, the algorithms proposed by Laporte, Nobert, & Desrochers [25,26] exist. In the second category, many heuristics and metaheuristics have been proposed (for a detailed survey of exact approximate algorithms for the DCVRP, see [46]).

2.2   Vehicle Routing Problem with Time Windows

The VRPTW is a variant of the CVRP adding the following constraints: (i) the service of each customer i starts within a time window $[e_i, l_i]$, and (ii) each vehicle v stops for s_i time instants in order to serve customer i. In case of early arrival to a customer i, the vehicle waits until the service time, i.e. until time instant e_i.

Exact and heuristic algorithms exist for the solution of the CVRP. In the first category, the algorithms proposed by Cook & Rich [9], Kallehauge, Larsen, Madsen [20], Kohl et al. [22], and Larsen [28]. In the second category, many heuristics [17] and metaheuristics [36,43] have been proposed (for a detailed survey of exact and approximate algorithms for the VRPTW, see [46]).

2.3 Vehicle Routing Problem with Backhauls

The VRPB is a variant of the CVRP where the N customers are partitioned into two subsets: (i) the first subset, L, contains l Linehaul customers that require a given quantity of products to be delivered, and (ii) the second subset, B, contains b Backhaul customers that require a given quantity to be picked up. The VRPB alters the third constraint of the CVRP, i.e. the total demand of each route must not exceed the capacity of each vehicle, in the following manner: the total demand of the linehaul and backhaul customers visited in a route do not exceed separately the vehicle capacity. Furthermore, the VRPB adds the following constraint: all deliveries must be made on each route before any pickup can be made.

Exact and heuristic algorithms exist for the solution of the CVRP. In the first category, the algorithms proposed by Mingozzi, Giorgi, & Baldacci [30], and Toth & Vigo [45]. In the second category, many heuristics have been proposed [6,11,15] (for a detailed survey of exact and approximate algorithms for the VRPB, see [46]).

2.4 Vehicle Routing Problem with Pickup and Delivery

The VRPPD is a variant of the CVRP where customers can also return a given quantity of products. Each customer i is associated with two quantities: (i) d_i denotes the demand of customer i, and (ii) p_i denotes the quantity of products that is picked up at customer i. Let us denote with O_i the node that is the origin of the delivery quantity and D_i the node that is the destination of the pickup quantity. These nodes, O_i and D_i may be the central depot; in this case, the customer will be served from products in stock or the products picked up at the customer will return to the central depot, respectively. If this is not the case, then a given quantity of products that are picked up from customer O_i can be delivered to customer i and a given quantity of products picked up at customer i can be delivered to customer D_i. The VRPPD alters the third constraint of the CVRP, i.e. the total demand of each route must not exceed the capacity of each vehicle, in the following manner: the current load of each vehicle along a route must be non-negative and must not exceed the capacity of the vehicle. Moreover, the VRPPD adds the following constraints: (i) Each customer O_i, when different from the central depot, must be served in the same route and before customer i, and (ii) Each customer D_i, when different from the central depot, must be served in the same route and after customer i.

Exact and heuristic algorithms exist for the solution of the VRPPD. In the first category, the algorithms proposed by Dumas, Desrosiers, & Soumis [12], Ruland and Rodin [38], and Savelsbergh & Sol [40]. In the second category, many heuristics have been proposed [33,49] (for a detailed survey of exact and approximate algorithms for the VRPB, see [46]).

3 Tangible User Interfaces

A tabletop is a horizontal multi-gesture user interface surface that provides co-location and interaction for single and multiple users [31] and is a useful shared space for diverse collaborative tasks [50]. Traditional input devices can be replaced by a tabletop display. By utilizing a tabletop display, users can interact with more natural devices. A tabletop can be handled by: (i) finger, (ii) hand gestures, and/or (iii) controller objects. In this paper, we use all these approaches; decision makers can specify customers' locations using controller objects and more specifically fiducials and can input/alter other parameters through finger and hand gestures.

Tabletop displays have been widely used in decision-making process. Kientz et al. [23] proposed a DSS to support collaborative decision-making for home-based therapy teams. Scotta et al. [42] presented a multi-user tangible interface system that aims at introducing an instrument to improve the response phase of the decision-making process. Hofstra et al. [16] used multi-user tangible interfaces for decision-making in disaster management. Scott et al. [41] have used tabletop interfaces to support collaborative decision-making in maritime operations. Engelbrecht et al. [13] used digital tabletops for situational awareness in emergency situations. Ploskas et al. [34] proposed an interactive spatial DSS with tangible user interfaces through a tabletop that supports decision-making and integrates geographical information data in the DSS for the Multiple Capacitated Facility Location Problem. The proposed paper uses the same tabletop interface with our previous work [34], but solves another problem (four CVRP variants) and uses different collaborative strategies. Finally, in our previous work [35], we proposed a collaborative spatial DSS for the CVRP on a tabletop display that allows two decision makers to collaborate with each other in order to find the best possible solution. This paper is an extension of our previous work [35] and incorporate in our spatial DSS four variants of the CVRP.

The tabletop used in this paper has been designed and constructed from scratch. The key design features are thoroughly described in [3].

4   Design, Implementation and Presentation of the Collaborative DSS

Figure 2 presents the decision making process that the decision makers can perform using the DSS. Initially, the decision makers select the type of the problem among five variants of the VRP: (i) CVRP, (ii) DVRP, (iii) VRPTW, (iv) VRPB, and (v) VRPPD. In the previous step, a small description of the five variants is presented in order for the decision makers to understand the implications of the selection. Then, the decision makers select the location of the central depot via an interactive Google Map using fiducials on the tabletop. In the next step, the decision makers select the location of the customers and for each customer input the model data (e.g. demand, service time, etc.) depending on the variant that they selected (Fig. 3). When the market is large and the manual integration of the data is not convenient, decision makers can download an Excel template, incorporate their data and upload the Excel file to the DSS. After this step the final representation of the problem is presented to the decision makers. Then, the tabletop display is divided into two segments, where each decision maker can input different model parameters (i.e. number of vehicles, vehicles' capacity, fixed cost, and cost per km) and find a solution (Fig. 4). The solution is visualized through a Google Map and the decision maker can export a detailed report as a pdf file. Furthermore, one decision maker can press the share button in order to copy his/her model parameters and solution to the other decision maker's display. When the share button is pressed, the current solution is saved as the current best possible one. Figures 3 and 4 show the process of the solution of a VRPTW; similar steps are followed on the other VRP variants.

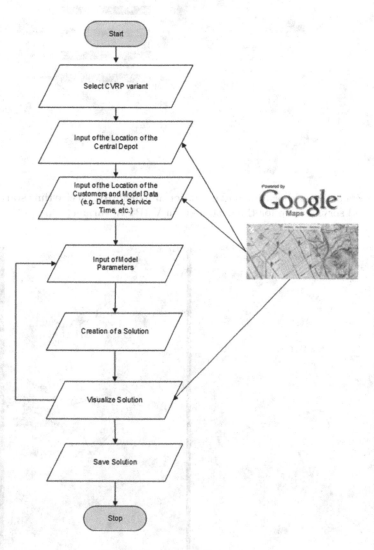

Fig. 2. Decision making process

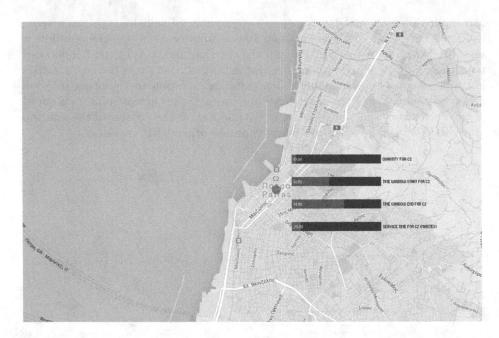

Fig. 3. Select customer's location and input demand quantity, service time start, service time end, and service time for the customer (in VRPTW variant)

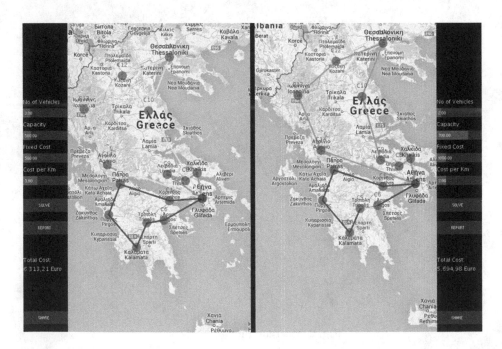

Fig. 4. Visualization of the solutions

The spatial DSS has been implemented using Java, TUIO, jsprit and Google Maps. More specifically, the open source TUIO protocol [21] has been used for the recognition of a set of objects with fiducials and draw gestures onto the table surface with the finger tips. Moreover, jsprit [19], a java based, open source toolkit for solving rich traveling salesman (TSP) and VRP variants, has been utilized in order to find a solution for the given problems.

Community Core Vision (CCV), previously known as tbeta, is an open source software that takes as input a video stream and outputs several tracking data, such as coordinates of the objects or events like finger down [8]. The recognition of the camera from CCV requires the installation of the device driver named CL-EYE Platform Driver. Moreover, open source Unfolding library [48] for Java was used to create interactive Google Maps and geovisualizations. The library supports various functions to get automatically the distance in km between two points in the earth.

5 Conclusions

The CVRP is the classical variant of the VRP, which is a well-known combinatorial optimization problem with many practical applications. In this paper, we extend our previous work [35] and propose a collaborative spatial DSS that allows two decision makers to collaborate with each other in order to find the best possible solution of a given CVRP and its' four well-known variants: (i) DVRP, (ii) VRPTW, (iii) VRPB, and (iv) VRPPD. The addition of these variants to the DSS allows decision makers to solve their routing problems by representing the real market and giving the actual constraints and not making any abstractions. Decision makers can initially locate the customers through interactive and user-friendly Google Maps and experiment with the problem's parameters to explore various solutions. Then, they can compare their solutions and find the best possible one. In future work, we plan to evaluate the proposed DSS and compare it with a standard non–tangible solution.

References

1. Anderson, D., Anderson, E., Lesh, N., Marks, J., Mirtich, B., Ratajczak, D., Ryall, K.: Human-guided simple search. In: Seventeenth National Conference on Artificial Intelligence Conference Proceedings (AAAI-2000), pp. 209–216 (2000)
2. ArcGIS: http://www.esri.com/software/arcgis. Last access on 10 April 2014
3. Athanasiadis, I.: Analysis, design and implementation of a hybrid system for teaching basic programming structures using traditional and tangible user interfaces to pre teenagers. MSc Thesis. School of Science and Technology, Hellenic Open University (2014)
4. Blasum, U., Hochstttler, W.: Application of the branch and cut method to the vehicle routing problem. Technical report, zaik2000-386, Zentrum fur Angewandte Informatik Koln (2000)
5. Bodin, L., Golden, B.L., Assad, A., Ball, M.O.: The state of the art in the routing and scheduling of vehicles and crews. Comput. Oper. Res. **10**, 69–221 (1983)

6. Clarke, G., Wright, J.V.: Scheduling of vehicles from a central depot to a number of delivery points. Oper. Res. **12**, 568–581 (1964)
7. Cordeau, J.F., Gendreau, M., Hertz, A., Laporte, G., Sormany, J.S.: New heuristics for the vehicle routing problem. In: Langevin, A., Riopel, D. (eds.) Logistics Systems: Design and Optimzation, pp. 279–297. Springer, New York (2004)
8. Community Core Vision: http://nuicode.com/projects/tbeta. Last access on 10 April 2014
9. Cook, W., Rich, J.L.: A parallel cutting plane algorithm for the vehicle routing problem with time windows. Technical report, Computational and Applied Mathematics, Rice University, Houston, TX (1999)
10. Dantzig, G.B., Ramser, R.H.: The truck dispatching problem. Manage. Sci. **6**, 80–91 (1959)
11. Deif, I., Bodin, L.D.: Extension of the clarke and wright algorithm for solving the vehicle routing problem with backhauling. In: Kidder, A. (ed.) Proceedings of the Babson College Conference on Software Uses in Transportation and Logistic Management, Babson Park, MA, pp. 75–96 (1984)
12. Dumas, Y., Desrosiers, J., Soumis, F.: The pickup and delivery problem with time windows. Eur. J. Oper. Res. **54**, 7–22 (1991)
13. Engelbrecht, A., Borges, M., Vivacqua, A.S.: Digital tabletops for situational awareness in emergency situations. In: 15th International Conference Proceedings on Computer Supported Cooperative Work in Design (CSCWD), pp. 669–676 (2011)
14. Fukasawa, R., Longo, H., Lysgaard, J., de Arago, M.P., Reis, M., Uchoa, E., Werneck, R.F.: Robust branch-and-cut-and-price for the capacitated vehicle routing problem. Math. Program. **106**(3), 491–511 (2006)
15. Goetschalckx, M., Jacobs-Blecha, C.: The vehicle routing problem with backhauls. Eur. J. Oper. Res. **42**, 39–51 (1989)
16. Hofstra, H., Scholten, H., Zlatanova, S., Scotta, A.: Multi-user tangible interfaces for effective decision-making in disaster management. In: Nayak, S., Zlatanova, S. (eds.) Remote Sensing and GIS Technologies for Monitoring and Prediction of Disasters, pp. 243–266. Springer, Heidelberg (2008)
17. Homberger, J., Gehring, H.: Two evolutionary metaheuristics for the vehicle routing problem with time windows. INFOR **37**, 297–318 (1999)
18. Ioannou, G., Kritikos, M.N., Prastacos, G.P.: Map-Route: a GIS-based decision support system for intra-city vehicle routing with time windows. J. Oper. Res. Soc. **3**, 842–854 (2002)
19. jsprit: https://github.com/jsprit/jsprit. Last access on 10 April 2014
20. Kallehauge, B., Larsen, J., Madsen, O.B.G.: Lagrangean duality and nondifferentiable optimization applied on routing with time windows-experimental results. Internal report IMM-REP-2000-8, Department of Mathematical Modelling, Technical University of Denmark, Lyngby, Denmark (2000)
21. Kaltenbrunner, M., Bovermann, T., Bencina, R., Costanza, E.: TUIO - a protocol for table-top tangible user interfaces. In: 6th International Workshop on Gesture in Human-Computer Interaction and Simulation Proceedings (2005)
22. Kohl, N., Desrosiers, J., Madsen, O.B.G., Solomon, M.M., Soumis, F.: 2-Path cuts for the vehicle routing problem with time windows. Transp. Sci. **33**, 101–116 (1999)
23. Kientz, J.A., Hayes, G.R., Abowd, G.D., Grinter, R.E.: From the war room to the living room: decision support for home-based therapy teams. In: 20th Anniversary Conference on Computer Supported Cooperative Work Proceedings, pp. 209–218 (2006)

24. Laporte, G.: Fifty years of vehicle routing. Transp. Sci. **43**(4), 408–416 (2009)
25. Laporte, G., Desrochers, M., Nobert, Y.: Two exact algorithms for the distance-constrained vehicle routing problem. Networks **14**, 161–172 (1984)
26. Laporte, G., Nobert, Y., Desrochers, M.: Optimal routing under capacity and distance restrictions. Oper. Res. **33**, 1050–1073 (1985)
27. Laporte, G., Semet, F.: Classical heuristics for the capacitated VRP. In: Toth, P., Vigo, D. (eds.) The Vehicle Routing Problem. SIAM Monographs on Discrete Mathematics and Applications, pp. 109–128. SIAM, Philadelphia (2002)
28. Larsen, J.: Parallellization of the vehicle routing problem with time windows. Ph.D. thesis. Department of Mathematical Modelling, Technical University of Denmark, Lyngby, Denmark (1999)
29. Lysgaard, J., Letchford, A.N., Eglese, R.W.: A new branch-and-cut algorithm for the capacitated vehicle routing problem. Math. Program. **100**(2), 423–445 (2004)
30. Mingozzi, A., Giorgi, S., Baldacci, R.: An exact method for the vehicle routing problem with backhauls. Transp. Sci. **33**, 315–329 (1999)
31. Madni, T.M., Sulaiman, S.B., Tahir, M.: Content-orientation for collaborative learning using tabletop surfaces. In: Proceedings of 2013 International Conference on Information Science and Applications (ICISA), pp. 1–6. IEEE (2013)
32. Naddef, D., Rinaldi, G.: Branch-and-cut algorithms for the capacitated VRP. In: Toth, P., Vigo, D. (eds.) The Vehicle Routing Problem. SIAM Monographs on Discrete Mathematics and Applications, pp. 53–84. SIAM, Philadelphia (2002)
33. Psaraftis, H.N.: k-Interchange procedures for local search in a precedence-constrained routing problem. Eur. J. Oper. Res. **13**, 391–402 (1983)
34. Ploskas, N., Athanasiadis, I., Papathanasiou, J., Samaras, N: An interactive spatial decision support system enabling co-located collaboration using tangible user interfaces for the multiple capacitated facility location problem. Int. J. Decis. Support Syst. Technol. **7**(2) (submitted for publication)
35. Ploskas, N., Athanasiadis, I., Papathanasiou, J., Samaras, N: A collaborative spatial decision support system for the capacitated vehicle routing problem on a tabletop display. In: Decision Support Systems IV - Information and Knowledge Management in Decision Processes. Springer Proceedings in Business Information Processing (submitted for publication)
36. Rochat, Y., Taillard, E.D.: Probabilistic diversification and intensification in local search for vehicle routing. J. Heuristics **1**, 147–167 (1995)
37. Ruiz, R., Maroto, C., Alcaraz, J.: A decision support system for a real vehicle routing problem. Eur. J. Oper. Res. **153**(3), 593–606 (2004)
38. Ruland, K.S., Rodin, E.Y.: The pickup and delivery problem: faces and branchand-cut algorithm. Comput. Math. Appl. **33**, 1–13 (1997)
39. Santos, L., Coutinho-Rodrigues, J., Antunes, C.H.: A web spatial decision support system for vehicle routing using Google Maps. Decis. Support Syst. **51**(1), 1–9 (2011)
40. Savelsbergh, M.W.P., Sol, M.: Drive: dynamic routing of independent vehicles. Oper. Res. **46**, 474–490 (1998)
41. Scott, S.D., Allavena, A., Cerar, K., Franck, G., Hazen, M., Shuter, T., Colliver, C.: Investigating tabletop interfaces to support collaborative decision-making in maritime operations. In: International Command and Control Research and Technology Symposium Proceedings (ICCRTS 2010), pp. 22–24 (2010)
42. Scotta, A., Pleizier, I.D., Scholten, H.J.: Tangible user interfaces in order to improve collaborative interactions and decision making. In: 25th Urban Data Management Symposium Proceedings, pp. 15–17 (2006)

43. Taillard, E.D., Badeau, P., Gendreau, M., Guertin, F., Potvin, J.Y.: A tabu search heuristic for the vehicle routing problem with soft time windows. Transp. Sci. **31**, 170–186 (1997)
44. Tarantilis, C.D., Kiranoudis, C.T.: Using a spatial decision support system for solving the vehicle routing problem. Inf. Manage. **39**(5), 359–375 (2002)
45. Toth, P., Vigo, D.: An exact algorithm for the vehicle routing problem with backhauls. Transp. Sci. **31**, 372–385 (1997)
46. Toth, P., Vigo, D.: The Vehicle Routing Problem. Siam, Philadelphia (2001)
47. Toth, P., Vigo, D.: Branch-and-bound algorithms for the capacitated VRP. In: Toth, P., Vigo, D. (eds.) The Vehicle Routing Problem. SIAM Monographs on Discrete Mathematics and Applications, pp. 29–51. SIAM, Philadelphia (2001)
48. Unfolding: http://unfoldingmaps.org/. Last access on 10 April 2014
49. Wilson, H., Sussman, J., Wang, H., Higonnet, B.: Scheduling algorithms for diala-ride systems. Technical report USL TR-70-13, Urban Systems Laboratory, MIT, Cambridge, MA (1971)
50. Yoshida, S., Yano, S., Ando, H.: Implementation of a tabletop 3D display based on light field reproduction. In: Posters of ACM SIGGRAPH 2010, p. 61. ACM (2010)

Decision Support Model for Participatory Management of Water Resource

Annielli Cunha[✉] and Danielle Morais

Production Engineering Department, Federal University of Pernambuco, Recife, Brazil
annielli.rangel@yahoo.com.br, dcmorais@ufpe.br

Abstract. Water resources are unevenly distributed across the territory and there is no stability concerning the availability of these resources. Conflicts over water use, particularly in situations where these resources are shared with other localities, require a participatory governance focused on cooperation and conflict resolution. For this reason, this study proposes a group decision model focusing on the use of the Problem Structuring Method (PSM) to assist decision-makers in identifying decision alternatives and evaluating them based on a voting system to concentrate the group decision around the relevant alternatives. Therefore, this group decision support model allows a full understanding of the problem by decision-makers and encourages the participation of those involved in a structured manner, reducing the negative consequences and dissatisfaction of those involved.

Keywords: Water conflicts · Decision support · Problems structuring methods

1 Introduction

Factors such as population growth, growth in agricultural and industrial production, unplanned urbanization, and poor management in addition to numerous other man-made activities, have increased the current demand for water. Additionally, water resources are distributed unevenly across territories and there is no stability concerning the availability of these resources.

Water conflicts have left their mark in history and, at times, were the cause of wars. The first documented war caused by competition for water occurred near the junction of the Tigris and Euphrates Rivers. It took place 4,500 years ago in the region currently known as Iraq. The armies of Lagash and Umma, battled with spears and chariots after Umma's king drained an irrigation canal leading from the Tigris [1].

Other conflicts that have marked history happened between China and Tibet [2], Syria and Turkey. [3] Israel was involved in a succession of water disputes with Syria, Lebanon and Jordan, as well as with the Palestinian [4].

Water management involves multiple uses and conflicting interests [5, 6]. Resources should meet the demands of animal consumption, factories, irrigation, transportation, power generation, and others, with human consumption as a priority. The decision concerning the use of water has increased its complexity when that resource is shared.

© Springer International Publishing Switzerland 2015
B. Delibašić et al. (Eds.): ICDSST 2015, LNBIP 216, pp. 85–97, 2015.
DOI: 10.1007/978-3-319-18533-0_8

In the case of water resources shared by different countries, states or municipalities, conflicts may be unavoidable.

Current water crises caused by climate change and depletion of existing resources have evoked the discussion of the possible and acceptable alternatives to managing the crisis, orchestrating the conflict, and defining the necessary decisions. The decision will usually be the product of an interaction between the individual preferences of all individuals (actors) involved [7], which can increase the complexity.

The use of models to support the decision-making process is discussed as an important tool in supporting the decision [8]. This is because models help to put the complexities and possible uncertainties that accompany the problem of decision-making within a logical structure subject to a comprehensive analysis, where one can understand the decision alternatives, the expected effects, and relevant data to analyze the alternatives and lead to conclusions [9].

This approach refers to a set of methods that seek a learning process and standardization of information about the issue between the parties involved [10]. The Problem Structuring Methods (PSM) have been developed to support the process of group decisions, allowing the actors to understand the focus of the problem and commit to a subsequent action. This allows for the different ways that the actors involved are used to reflect upon future decisions and suggest options for the resolution of issues considered complex for decisions in an environment of uncertainty and conflict [11].

On the other hand, by using PSM facilitates to understand the problem, it is sometimes hard to reach a compromise solution among those achieved by PSM, considering the plurality of divergent interests among decision makers in the management of shared water resources. Thus, in order to improve this situation, a voting system can be applied to aggregate the preferences and find a solution where all participants are involved. There have been studies where voting processes was used to achieve a group decision in water resources context [12–14].

Therefore, the goal of this study is to propose a model consisting of two main phases to support group decision making: alternatives identification and voting system. The first phase uses a PSM, allowing a complete understanding of the issue by the decision-makers and generating alternatives. The second phase applies a voting system to evaluate the actions to be implemented through the aggregation of individual preferences. This model encourages the participation of decision-makers in a structured manner and then reduces the negative consequences and dissatisfaction of those involved.

The following section presents an overview of decision support process in water management, focusing on managing conflicts around issues on the use of shared water. Next, the proposed model is discussed and a real situation of the conflict in the management of a watershed that crosses three states in Brazil will be presented, stressing how the proposed model could help identify solutions. Finally, the conclusions and limitations of this work will be in the last section of this paper.

2   Decision Support Models in Water Planning

Improving the decision-making process is a challenge that has led to the development of several approaches to decision support.

Decision-making analyses that are capable of conflict resolution are particularly useful tools in analyzing decision problems that extend to the level of accommodating the stakeholders' preferences [15].

Factors such as climate change and increasing demand for water have motivated discussions on water management. At the same time, the amount of research and number of publications on water conflicts and decision support tools have increased. Some publications are a result of major international groups dedicated to research in water management areas, such as the IIASA (International Institute for Applied Systems Analysis), founded in 1972 [16].

Some important publications that integrate the use of decision support tools with shared river basins problem are described below:

A participatory decision model is used to facilitate science and political integration when using hydrological models and evaluating scenarios in the Jordan River water management [17]. China shares 110 rivers and lakes with 18 downstream countries.

In [18] hydrology is applied as a mechanism to inform the development of a transboundary environmental compensation mechanism and regional consultative mechanisms that support informed, cooperative decision-making for China and its riparian neighbors. The importance of co-operative development and coordinated management of international rivers in western China is highlighted in [19].

In [20] the known method of multi-criteria decision Analytic Hierarchy Process (AHP) was used to model the optimal water distribution Djerdap I. It is a hydro-electric power plant built as a concrete dam across the transboundary waters of the Danube between Serbia and Romania. A dynamic system was developed in [21] and its usefulness was evaluated in the modeling of complex water systems using the Bear River Basin as a data source. The transboundary basin includes portions of Idaho, Utah, and Wyoming.

To analyze the consequences of various policy alternatives in an international river (the Ganges), a prototype Web-based Decision Support System (DSS) was developed. The application of the system enabled the identification of the potential of this approach in the systematic management of shared river basins, leading to effective conflict resolution [22].

The importance of public participation and support, as well as the identification of appropriate strategies and political commitments of key decision makers, is discussed in the study evaluating the diagnosis that preceded the action plan formulated for the San Juan River basin, shared by Costa Rica and Nicaragua in Central America [23].

The political debate on transboundary waters is enriched in [24], by trying to demonstrate the effectiveness of a critical hydropolitical approach, since the literature is limited, especially on international river basins. An important contribution to the shared water management was the development of an integrated decision support process to achieve a consensus. The participatory decision-making system consists of three phases: define and structure the problem, identify possible alternatives, and formulate the way that participants judge the problem and achieving consensus. PSM are used to support the stakeholders in providing their perspective of the problem and to elicit their interests and preferences [25].

3   Proposed Decision Support Model

The PSM are indicated to improve the understanding of the problem by all those involved and ensure that the decision process is conducted in the most appropriate way. Thus, PSM are defined as a family of decision support methods that helps groups of diverse interests reach a consensus concerning the problem and commit to a consequent action [26].

Some of the advantages in using the PSM include:

– helping participants of a decision-making process to better understand their priorities, justifying them properly, legitimizing their findings, and facilitating the validation of the decision-making process [27],
– accommodating multiple alternative perspectives, facilitating the negotiation of a joint agenda, functioning through interaction and iteration, and generating ownership of the modeling of the problem and its action implications through transparency of representation [28].

PSM are used in water resource management problems because they allow the improvement of communication between stakeholders as well as the organizing and synthesizing of information so that knowledge about the problem and the scope of the decision is expressed and understood by all.

The use of tools that help in the mediation of conflict, improve communication between the parties involved, and support the search for better alternatives are indicated in the decision on the use of shared river basins, especially in the context of water crisis, where conflicts are inherent in the decision-making process. For this, a multi methodology is used, applying SSM and a voting system. Figure 1 illustrates the proposed method.

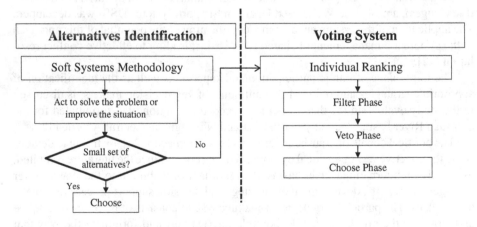

Fig. 1. Phases of the proposed model

The Soft Systems Methodology (SSM) stands out among the methods that fit the purpose of generating alternatives. The choice for SSM is justified because the method uses a sequence of steps that generates adequate alternatives, based on what is known about the problem and what is possible to implement.

This proposed group decision model contributes to improving the water management process since it makes the decision making process more participative.

3.1 Alternatives Identification

Soft Systems Methodology (SSM) is designed to help formulate and solve situations described as "soft". These unstructured and complex problems usually involve several human components, presenting, according to this characteristic, different perceptions of the same problem or goal, or/and different Weltanschauungen (worldviews) of the various stakeholders involved in the system. Learning occurs in SSM by comparing pure models of objective activity (in the form of system models of human activity) to the perceptions of what action is happening in the real world in a problematic situation.

For this, the methodology is divided into seven stages as show in Fig. 2

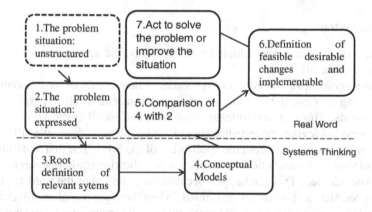

Fig. 2. Representation of the stages of SSM. Adapted from [24].

In stages 1 and 2, the information is captured and analyzed. For this, a cognitive and visual representation mechanism, called rich figure, is used. The collection of the interests of stakeholders can be done through interviews, workshops, discussions, data collection, etc. Use of the rich figure seeks to define the problem and a shared expression of the situation enables a common understanding of the problem.

In stage 3 the expressions of the problem situation, obtained in the previous stage, are useful to initiate the thinking about the situation and for defining the problem. Root definitions describe fundamental characteristics of systems organized about important issues or major tasks relevant to the problem situation. A common technique to initiate the process of conceptual modeling in SSM is the development of root definitions of human activity systems.

In stage 4 each root definition is expanded into a conceptual model, defining the activities necessary for the business to meet the purpose specified and indicating relationships among the activities.

In stage 5, there is a comparison between the conceptual model developed in stage 4 and the actual situation extracted from the Rich Picture on stage 2. This comparison structure allows for a discussion about the differences identified between models. The discussion addresses what is missing in the actual situation that comes close to the conceptual model [29].

In stage 6, suggestions extracted in the previous stage are discussed and the feasibility and acceptability is verified. Finally in stage 7, the action to improve the problem situation should be implemented.

This approach helps build models of action to solve problems and seeks to discover the crucial aspects of a problem situation, encouraging people to think about how this problem arose or what their developers' processes were. It is widely used as a promoter of consensus, as each individual can see the influence caused by their own mental model [30].

The result of the SSM is the definition of possible and desirable alternatives to solve the problem, or the identification of the main agreements and disagreements among the participants.

3.2 Voting System

The definition of the alternative adopted will depend on the aggregation of individual preferences of those involved.

In the management of shared water resources, where conflicts are inherent in the decision-making process, the expectation of a consensus can result in periods of time without decisions. This can lead to legal, economic, and social costs. Thus, the group decision process achieving convergence should not be the main objective [19].

Two situations may result from the application of the SSM. The first situation is the generation of a set of succinct alternatives or even an indication of alternatives or courses of action to be adopted. This can happen because the potential of the SSM application orchestrates conflicts and promotes consensus. When there are few alternatives and the processes of learning and discussion did not result in a consensus, negotiation procedures are required.

The second situation is the identification of a large number of creative alternatives. In this case, it is recommended to use voting methods.

The weighted voting method by Quartile, described in [14] was chosen because it allows a large number of members to be involved in the process and considers that each stakeholder may interpret the problem differently. Thus, each participant makes a ranking of the alternatives of a common group individually. Additionally, the method considers the need for participants to understand the process of aggregation. Increased transparency of the method can generate greater trust and acceptance. The weighted voting method by Quartile has three phases: filter, veto, and choose. Using the individual ranking of the alternatives generated by members of a group in the filter phase, two sets of alternatives are created, considered an upper and lower order, applying the separation quartile.

The amount of times that i appear in the upper quartile is equal to the number of votes that an alternative receives. This amount is represented by Q_i^u. The alternatives that have not been considered by any of the makers, that is, $Q_i^u = 0$, must be eliminated.

The second filter is an analysis of the alternatives that make up the lower quartile. The amount of times an alternative appears in the lower quartile is called Q_i^L and represents the number of votes that the alternative i has to be among the worst. So if an alternative receives more votes against than in favor, this is, if $Q_i^L \geq Q_i^u$, the alternative i is eliminated.

Thus, it is possible to eliminate dominated alternatives and focus efforts to decide among the relevant alternatives.

In the veto phase, a positional count of the alternatives is used based on Borda's method. However, only the alternatives found in the upper and lower quartiles are analyzed. The veto phase eliminates alternatives not approved by most makers thereby arriving at a set of alternative accepted by the participants.

It is possible to identify whether there is a strong opposition to the alternative selected in the second filtering. For this step, you can eliminate the alternatives classified as worse for most decision-makers by the strength and weakness analysis of the alternatives. Strength is determined by Eq. 1

$$F_i = \sum_{k=1}^{m} \sum_{j=1}^{n/4} \left(\frac{n}{4} - j + 1 \right) q_{ij}^k \tag{1}$$

Weakness analysis of the alternatives is determined by Eq. 2.

$$f_i = \sum_{k=1}^{m} \sum_{\left(\frac{3n}{4}+1\right)}^{n} \left(j - \frac{3n}{4} \right) q_{ij}^k \tag{2}$$

Where i are the alternatives, j is the position in the ranking, k represents the decision makers that range from 1 to m, and q is the number of votes.

If $f_i \geq F_i$ then i should be eliminated.

It should also evaluate whether there is any opposition to the alternative selected. The procedure ends when it identifies the intensity of the alternative's strength. This occurs through the expression $\alpha_i = F_i - f_i$. The alternative chosen is the one with higher value α_i.

Finally, the choice phase occurs when the alternative with the highest score is selected.

When understanding and learning about the problem have provided an alignment that allows decision-makers to choose an alternative at the end of the process, perhaps due to very intense divergence, the method for trading should be used to analyze preferred alternatives and seek agreements among decision makers.

4 Water Crisis in Brazilian Southeast

The southeastern region of Brazil is facing the greatest water crisis of all time. The region is made up of three states: Minas Gerais, Rio de Janeiro and São Paulo, where the latter is the state most affected by the crisis.

The state of São Paulo has the largest industrial park in Brazil. It accounts for approximately 30 % of national GDP, the largest in the country. The state population is estimated at more than 44 million people, according to the most recent census.

The eight supply systems have declined in 2014 due to drought and poor planning. Meanwhile, the population suffers from supply disruptions and companies are forced to interrupt their industrial activities. This affects the state of growth and, considering the great economic contribution of the state, the economic growth of the country.

The Paraíba do Sul River is of strategic importance for the three states and supplies water to 184 cities. The river is born in São Paulo and goes through a state of the Minas Gerais and culminates in Rio de Janeiro. The transposition of the Paraíba do Sul is a government project from Sao Paulo to divert water to its main supply system.

This project resulted in a federal conflict with the other states of the Southeast. The Federal Public Ministry of Rio de Janeiro filed a civil suit in federal court asking that an injunction be granted to block the state government of Sao Paulo project.

Due to the seriousness of the situation, the Department of Justice declared that a public hearing was necessary. At this meeting, the states expressed a mutual desire to help solve the problem and signed an agreement pledging not to carry out works without the consent of all parties involved. The three states agreed to submit, after three months, proposals to combat the water crisis. An illustrative example is presented to this problem.

4.1 Illustrative Example of the Proposed Model

The application of SSM involved 6 representatives, listed in Table 1.

Table 1. Decision Makers

Decision Maker	Representation	Decision Maker	Representation
DM1	Government of Sao Paulo	DM4	Federal Agency
DM2	Government of Minas Gerais	DM5	Technical Committee
DM3	Government of Rio de Janeiro	DM6	Environmental Committee

The application of SSM allows the problem to be precisely defined and constraints specified. Perceptions, goals, and views of all the decision makers were addressed during the exploration of the collective understanding about the problem.

The first phase of the model ended with a set of feasible alternatives that were considered acceptable by comparing the actual and the desired position obtained by the conceptual model.

This resulted in the generation of 12 alternatives to solve the problem of the water crisis. Only one alternative should be chosen as a priority action for implementation. Decision-makers assess the alternatives differently because they have different parameters of judgment and priority (immediate changes, long-term changes, financial costs, social, political image, etc.). Table 2 shows these alternatives.

Table 2. Alternatives description

Code	Description	Code	Description
A	Transposition of the river Paraíba do Sul	G	Construction of desalination plants
B	Construction of new reservoirs	H	Treatment and reuse of wastewater
C	Establish obligation to construct rainwater-collection devices	I	Air collectors to condense water
D	Establish mandatory building of artesian wells in all new buildings	J	Changing cultural habits
E	Use of underground reservoir Guarani	K	Loss Reduction
F	Depollution of the rivers Tietê and Pinherinhos	L	Rotation of water supply

The voting system allows each decision maker to evaluate the alternatives according to their goals. Table 3 shows the order in which the alternatives were evaluated by each decision maker. Each decision maker evaluated a different alternative in the first position of the ranking, which demonstrates the differences between them.

Table 3. Individual Rankings

Ranking	DM1	DM2	DM3	DM4	DM5	DM6
1ª	A	F	J	H	B	K
2ª	B	C	D	C	H	J
3ª	F	D	F	D	C	H
4ª	E	K	K	B	F	C
5ª	C	H	C	E	E	F
6ª	H	J	G	F	A	G
7ª	D	I	E	K	D	B
8ª	K	B	H	A	G	D
9ª	I	A	I	G	J	E
10ª	G	E	A	I	K	I
11ª	L	G	B	L	I	A
12ª	J	L	L	J	L	L

The first phase consists of analyzing alternatives which are in the upper and lower quartiles. As the total number of alternatives is n = 12, then the first three alternatives are in the upper quartile and the last three alternatives are in the lowest quartile.

The number of times that an alternative appears in the upper quartile, is the number of votes that this alternative has to be among the best. The Table 4 shows the number of votes that each alternative has in the upper quartile.

Table 4. Upper order

$Q_A^U = 1$	$Q_D^U = 3$	$Q_G^U = 0$	$Q_J^U = 2$
$Q_B^U = 2$	$Q_E^U = 0$	$Q_H^U = 3$	$Q_K^U = 1$
$Q_C^U = 3$	$Q_F^U = 3$	$Q_I^U = 0$	$Q_L^U = 0$

The first filter eliminates the alternatives that were not considered among the best for any decision maker. Thus, the alternative E, G, I and L are eliminated since $Q_i^U = 0$.

Analysis of the bottom quartile allows us to identify the number of votes that an evaluated alternative has among the worst. The number of votes of the worst alternative is shown in the Table 5.

Table 5. Lower Order

$Q_A^L = 2$	$Q_C^L = 0$	$Q_F^L = 0$	$Q_J^L = 2$
$Q_B^L = 1$	$Q_D^L = 0$	$Q_H^L = 0$	$Q_K^L = 1$

If an alternative has more votes against (or the same) than in favor, $Q_i^L \geq Q_i^U$ this alternative i must be eliminated. Thus, alternative A, J and K are eliminated.

In the veto phase, Eq. 1 is used to identify the strength of an alternative. In turn, Eq. 2 is used to calculate the alternative's weakness. This makes it possible to calculate the intensity of force for each alternative that has passed through the filter stage. Table 6 shows the results.

Table 6. Strength, weakness, and intensity of the force of the alternatives

Strength	$F_B = 5$	$F_C = 5$	$F_D = 4$	$F_F = 5$	$F_H = 6$
Weakness	$f_B = 2$	$f_C = 0$	$f_D = 0$	$f_F = 0$	$f_H = 0$
Intensity	$\alpha_B = 3$	$\alpha_C = 5$	$\alpha_D = 4$	$\alpha_F = 5$	$\alpha_H = 6$

Alternative H should be chosen because it has greatest force intensity. Thus, the shares relating to Treatment and Sewage Water Reuse should not implemented.

This case illustrates a water conflict situation, due to the existence of shared basins where agreements are needed. The urgency and essentiality of the resource highlight the need of mediation mechanisms that will lead to effective solutions and agreements among stakeholders.

The proposed model has the potential to aid in the decision making process by allowing participants to reflect on their goals, interests, priorities, and share with each other their view of the problem. Thus, participants can improve their understanding of the problem and the factors linked to them and find appropriate solutions to the problem of experienced water crisis.

5 Conclusions

Conflicts involving the use of shared water resources have a historical impact. The chance of these conflicts increasing due to changes in demand and availability is imminent.

The use of water is vital to maintaining quality of life and economic development, and decisions on its use need to be made considering the balancing of multiple criteria and various stakeholders.

This paper presented a model to support a group decision making process allowing participants to interact in the search for more appropriate collective solutions to real needs. For this, the SSM is recommended to identify alternatives. The method considers the current conditions of the problem and the expectations and interests of the participants.

The problem of a structuring method is followed by a voting system that has the potential to aggregate individual opinions, even divergent ones, and reach an alternative decision that addresses the collective interest.

In the illustrative example, the used of the voting system is presented and the result indicated that it is possible to add different individual reviews and find a possible solution approved by all participants.

In the management of water resource shared the use of a decision model has how potential benefits: orchestrate conflicts, enable a process of participatory decision-making, and reduce dissatisfaction and negative consequences.

The application of the model in real situations is identified as suggestions for continuity of work.

Acknowledgments. This study is part of a research program funded by the Brazilian Research Council (CNPq), to whom the authors are grateful.

References

1. Hammer, J.: Scarce tactics: killings in Iraq. Fighting in Syria. Is a lack of water to blame? Smithsonian **44**(3), 18 (2013)
2. Sullivan, P.: Glaciers, monsoons, rivers, and conflict: China and South Sea. Georgetown J. Int. Aff. **12**(1), 107 (2011)
3. Hipel, K., Kilgour, D.M., Kinsara, R.: Strategic investigations of water conflicts in the middle east. Group Decis. Negot. **23**(3), 355–376 (2014)

4. Venter, Al J.: The oldest threat: water in the Middle East. Middle East Policy. **6**(1), 126 (1988)
5. Morais, D.C., Almeida, A.T.: Water supply system decision making using multicriteria analysis. Water SA **32**(2), 229–235 (2006)
6. Fontana, M.E., Morais, D.C.: Using promethee V to select alternatives so as to rehabilitate water supply network with detected leaks. Water Resour. Manag. **27**(11), 4021–4037 (2013)
7. Figueira, J., Greco, S., Ehrgott, M.: Multiple Criteria Decision Analyses: State of the Art Surveys. Springer Science + Business Media Inc., Boston (2005)
8. Cunha, A., Morais, D.: Proposed multicriteria model for group decision support in water resources planning. seoul. In: EEE International Conference on Systems, Man and Cybernetics (2012)
9. Harvey, M.W.: Operations Research (Pesquisa Operacional). Prentice-Hall, Englewood (1986)
10. Cunha, A., Morais, D.: Analysis of problem structuring methods to improve decisions in environmental planning. In: 2014 IEEE International Conference on Systems, Man and Cybernetics (SMC), pp. 289–294 (2014)
11. Almeida, A., Morais, D., Costa, A.P., Alencar, L., Daher, S.: Decisão em Grupo e Negociação: métodos e aplicações (Group Decision and Negotiation: Methods and Application). Atlas, São Paulo (2012)
12. Trojan, F., Morais, D.: Prioritizing alternatives for maintenance of water distribution networks: a group decision approach. Source: Water SA **38**(4), 555–564 (2012)
13. Morais, D., Almeida, A.: Water network rehabilitation: A group decision-making approach. Source: Water SA **36**(4), 487–493 (2010)
14. Morais, D., Almeida, A.: Group decision making on water resources based on analysis of individual rankings. Omega **40**, 42–52 (2012)
15. Roozbahani, A., Zahraie, B., Tabesh, M.: PROMETHEE with precedence order in the criteria (PPOC) as a new group decision making aid: an application in urban water supply management. Water Resour. Manag. **26**(12), 3581–3599 (2012)
16. Kindler, J., Loucks, D.P.: Water resources research ai IIASA: 1973-1988. Water Resour. Manag. **3**, 169–190 (1989)
17. Comair, G.F., McKinney, D.C., Maidment, D.R., Espinoza, G., Sangiredy, H., Fayad, A., Salas, F.R.: Hydrology of the jordan river basin: a GIS-based system to better guide water resources management and decision making. Water Resour. Manag. **28**(4), 933–946 (2014)
18. He, D., Wu, R., Feng, Y., Li, Y., Ding, C., Wang, W., Yu, D.W.: China's transboundary waters: new paradigms for water and ecological security through applied ecology. J. Appl. Ecol. **51**(5), 1159–1168 (2014)
19. He, D.-M., Liu, X.-J., He, D.-M.: Equitable utilisation and effective protection of sharing transboundary water resources: international rivers of western China. Journal of Geographical Sciences. Vol.11(4) (2001)
20. Srdjevic, Z., Srdjevic, B.: Modelling Multicriteria decision making process for sharing benefits from the reservoir at serbia-romania border. Water Resour. Manag. **28**(12), 4001–4018 (2014)
21. Sehlke, G., Jacobson, J.J.: System dynamics modeling of transboundary systems: the bear river basin model. Ground Water **43**(5), 722–730 (2005)
22. Salewicza, K.A., Nakayama, M.: Development of a web-based decision support system (DSS) for managing large international rivers. Glob. Environ. Change **14**, 25–37 (2004)
23. Yamaguchi, H., Futamura, H., Nakayama, M.: Issues concerning a diagnostic study of an action plan for the San Juan river basin. Hydrological Process. **18**, 2977–2989 (2004)
24. Sneddon, C., Fox, C.: Rethinking transboundary waters: a critical hydropolitics of the Mekong basin. Political Geogr. **25**, 181–202 (2006)

25. Giordano, R., Passarela, G., Iricchio, V.F., Vurro, M.: Integrating conflict analysis and consensus reaching in a decision support system for water resource management. J. Environ. Manag. **84**, 213–228 (2007)

26. Franco, L.A., Cushmanb, M., Rosenhead, M.: Project review and learning in the construction industry: Embedding a problem structuring method within a partnership context. Eur. J. Oper. Res. **152**(3), 586–601 (2004)

27. Bouysoou, D., Marchant, T., Pirlot, M., Tarikias, A.: Evaluation and Decision Models with Multiple Criteria – Stepping Stones for the Analyst. Operational Research. Springer, Berlin (2006)

28. Rosenhead, M.: the Problema? An introduction to problem structuring method. Interfaces **26**(6), 117–131 (1996)

29. Cunha, A., Morais, D.: Drawing up a national plan for public sanitation: a participatory group decision approach. In: IEEE International Conference on Systems, Man, and Cybernetics, pp. 32–37 (2013)

30. Checkland, P.: Soft systems methodology. In: Rosenhead, J., Mingers, J. (eds.) Rational Analysis for a Problematic World Resisited, pp. 61–90. Wiley, Chichester (2004)

Modeling Interactions Among Criteria in MCDM Methods: A Review

Ksenija Mandic[✉], Vjekoslav Bobar, and Boris Delibašić

Faculty of Organizational Sciences, University of Belgrade,
Jove Ilica 154, Belgrade, Serbia
ksenija.mandic@crony.rs, vbobar@gmail.com,
boris.delibasic@fon.bg.ac.rs

Abstract. This paper reviews approaches for modelling interactions and dependencies between criteria in multi-criteria decision-making (MCDM) methods. Traditionally, MCDM methods only allow the establishment of linear dependence between criteria, so they only allow for simplified models that are mostly inadequate for modelling real-life problems. Several methods have therefore been developed for modelling the interdependence between criteria and sub-criteria in MCDM. This paper makes a comparison between some popular methods that allow modelling of criteria interdependencies: Analytic Network Process (ANP), Decision Making Trial and Evaluation Laboratory (DEMATEL), Interpretive Structural Modelling (ISM), Fuzzy Measures and the Choquet Integral (CI) and Interpolative Boolean Algebra (IBA). These methods allow the establishment of interactions and comparisons between criteria using supermatrices, diagrams, fuzzy measures, fuzzy integrals and logical functions. This paper presents the MCDM approaches that include in their analysis the interdependencies and relation between criteria/sub-criteria and thus enable more efficient and realistic modelling of decision-making problems.

Keywords: MCDM · Interactions between criteria · ANP · DEMATEL · ISM · Fuzzy measures and the Choquet Integral · IBA

1 Introduction

Multi-criteria decision-making (MCDM) is probably the most popular decision-making discipline. It models decision-makers' subjective assessments of a large number of quantitative and qualitative criteria, which are often conflicting. The primary aim of MCDM is to develop a methodology that enables the aggregation of criteria/sub-criteria, which includes the preferences of decision-makers [46]. Achieving this goal requires the application of complex procedures. As MCDM generally handles a multitude of criteria and sub-criteria, most literature on the topic uses the weighted sum of criteria for criteria aggregation. It is therefore assumed that criteria are independent, which is most often not the case. On the other hand, there are several approaches found in the literature that model interrelationship between criteria.

© Springer International Publishing Switzerland 2015
B. Delibašić et al. (Eds.): ICDSST 2015, LNBIP 216, pp. 98–109, 2015.
DOI: 10.1007/978-3-319-18533-0_9

When evaluating a decision-making problem, it is necessary to take into account a large number of criteria/sub-criteria and determine their relative weights. The criteria are often interdependent and between them there are certain relations, so their individual weight is hard to determine. For this reason, many decision-making problems cannot be adequately translated into a flat or even hierarchical criteria structure. In addition, the artificial neglect of these interdependencies affects the obtained results, so that they do not reflect the problem realistically. Therefore, in order to make an accurate and flexible decision, it is necessary to include in the MCDM analysis the interactions between the decision-making criteria. Engaging the complex relationships between the elements of a problem certainly requires more time and effort, but provides more realistic results.

One of the most commonly used MCDM methods is the Analytic Hierarchy Process (AHP) [31]. This technique compares and evaluates the impact of various elements in relation to the goal. It is based on a hierarchical, yet linear, structure between criteria. This method is not adequate, however, for the representation of the problem in a case where there are various interactions between elements, because it only takes into account a one-way hierarchical relationship between decision-making levels. In such situations, when decomposing the problem into a hierarchy, significant interdependence between the elements can be lost. Therefore, such cases require a holistic approach. In the literature, several methods have been developed that have tried to solve the discrepancy that emerges as a consequence of the failure to include interrelations between criteria in the decision-making process.

Analytic Network Process (ANP) was proposed to allow AHP [30] to model the interrelation between various hierarchical levels of decision-making and criteria. The hierarchical structures that are inherent to AHP are replaced by networks, within which relations between the levels are not represented in the manner of higher/lower, dominant/subordinate, or direct/indirect [23]. ANP is a non-linear structure that handles dependencies within a cluster of criteria (internal dependence) and between different clusters (external dependence) [5].

Another effective way to establish dependencies between decision-making criteria is proposed by the Decision-Making Trial and Evaluation Laboratory (DEMATEL) method. It was originally created between 1972 and 1979 by the Science and Human Affairs Program of the Battelle Memorial Institute of Geneva [11], in order to study complex and intertwined groups. DEMATEL visualizes complicated structural and causal relationships using matrices or digraphs and has the capability to convert relationships of cause and effect between criteria into a unique structural model [9].

Another approach applied in order to present interrelations among multiple variables is Interpretive Structural Modelling (ISM). This technique represents an interactive learning process, within which a set of various directly or indirectly related criteria is structured in a comprehensive systematic model [32]. ISM is a computer-aided method for developing graphical representations of system composition and structure [3].

In addition to the previously mentioned methods, it is recognized that fuzzy measures and integrals can also model interactions between criteria in a certain way [35]. It was not formalized, however, until Murofushi & Soned [25] proposed an interaction index for a pair of criteria. Later, Grabish [14] proposed a generalization of the index to any subset of criteria. At the beginning of the 90 s, the Sugeno integral was used as a tool for aggregation to calculate the average global score, taking into account the importance of criteria expressed by a fuzzy measure [14]. Then, after the proposal of Murofushi & Sugeno [25], the application of the Choquet integral (CI) - an extension of the classical Lebesgue integral - was quickly entered into use.

A more recent approach, which is singled out as suitable for representation of logical interactions between the criteria/sub-criteria of decision-making, is Interpolative Boolean Algebra (IBA). It was proposed by Radojevic [29] as a consistent realization of fuzzy logic [44]. When using IBA, all the axioms and theorems of Boolean logic apply [28]. What makes this model more flexible for application is that all the structural transformations are taken into account before assigning numerical values, which is not the case in conventional fuzzy MCDM methods. In addition, the IBA approach treats contradiction differently (i.e. a negated variable is not transformed immediately into a value) and respects the law of the excluded middle. Therefore, using IBA allows the establishment of fuzzy logic in a Boolean frame [27].

This paper is intended to review the aforementioned MCDM methods which can handle the relationships and interdependencies between decision-making criteria. Section 2 contains a literature overview of the MCDM methods which take into account the interactions between elements when making a decision. Section 3 gives a brief description of each of the presented methods. Sections 4 and 5 provide discussion and concluding observations with possible directions for further research.

2   Literature Review

Keeney & Raiffa [19] were among the first authors to analyse the problem of interactions between attributes within multi-attribute utility theory (MAUT). They considered three model variants. These models include as a key term the sum of the weighted attribute utilities. The most general model proposed is the multi-linear model, which includes the sum of weighted attributes and interaction terms. The multiplicative model is derived from the multi-linear model by setting all coefficients to one constant. The additive aggregation model is also a derivative of the multi-linear model, by setting all interaction coefficients as equal to zero. All three model variants are additive in essence.

In the literature, we can find several review papers of recent date which elaborate the methods of MCDM [1, 39, 40]. However, these papers are mainly based on a review of all the MCDM approaches, as well as their application in various fields of research. This review paper differs from the aforementioned papers in its analysis and description only of the methods primarily involving relationships and dependencies between criteria. Authors often combine the aforementioned MCDM methods in order to take into account the mutual interactions and interdependencies between the criteria/sub-criteria. Some combinations of these methods are presented hereinafter.

Gürbüz & Albayrak [16] propose a hybrid approach that combines ANP and CI for evaluation of human resources. The interaction between different criteria is taken into consideration, which is not peculiar for methods that have so far been used for the evaluation of human resources. In addition, two different types of interaction are managed at the same time. The reason for using ANP is that the decision-making problem has several criteria, and these criteria demonstrate interdependencies most of the time. On the other hand, CI - a fuzzy integral - handles "conjunctive/disjunctive" interactions between criteria. The same combination of methods is applied to the ERP selection problem [17].

On the other hand, Nguyen et al. [26] developed a fuzzy MADM model and machine tool selection, taking into account the interaction between criteria. For the needs of this paper they used fuzzy ANP and COPRAS-G (Complex Proportional Assessment of Alternatives with Grey Relations). The FANP is used to cope with imprecise information arising from the evaluations of decision-makers. Furthermore, this method allows the modelling of interaction, feedback, relationships and interdependence between criteria, and thus determines the weights of criteria. COPRAS-G enables the representation of preference ratios for the alternative interval values in relation to each criterion and to calculate the weighted priorities of the machine alternatives.

Mehregan et al. [24] studied the interaction between sustainability criteria in the selection of suppliers and to do so they used ISM and fuzzy DEMATEL (FDEMATEL) methods for the first time. They applied ISM to determine the interaction between sustainable supplier evaluation criteria, while the use of FDEMATEL allowed them to determine the intensity of these interdependencies. They illustrated how the integrated ISM-FDEMATEL model can be a significant management tool for evaluating and analyzing interactions between criteria. This approach has been used in Iranian gas engineering and in the development of companies.

Tadic et al. [38] have proposed a FDEMATEL-FANP-FVIKOR model for the selection of city logistics concepts. For the analysis of interdependences between factors and criteria fuzzy ANP was used, for determining interdependences between the groups of factors the fuzzy DEMATEL method was used, while the ranking of alternatives was done by FVIKOR technique. In the first phase, FDEMATEL and FANP are combined, providing a weight for each criterion. These weights are used in the next step, applying the FVIKOR method to rank the alternatives.

Another more recent approach that is singled out as an efficient tool for establishing a logical interaction between criteria/sub-criteria is IBA. Mandic et al. [22] have combined IBA and the classical MCDM method TOPSIS to conduct the selection of suppliers. IBA enables the presentation of logical dependencies between criteria with the help of Boolean algebra and in compliance with Boolean laws. In this study IBA has been used in order to present logical interdependences between the elements of decision-making, while TOPSIS was used to produce the ranking of alternatives.

The main criterion for the selection of sources cited within this section is the application of MCDM methods that involve interactions between criteria for solving real problems of decision-making. In the paper, works of recent date are quoted to indicate the topicality of analyzed themes.

3   Description of the Methods that Include Interaction Between the Elements of Decision-Making

3.1   Analytic Network Process - ANP

MCDM methods such as AHP and ANP are able to generate priority weights of criteria and alternatives, using a pair-wise comparison matrix of expert's decisions. However, ANP [30] is an expansion of AHP because it takes into account the relationships between higher-level and lower-level elements. ANP is used to model the interaction, dependence and feedback within groups of elements and between groups [26]. The groups of elements include the goals, criteria and sub-criteria in the decision making process. ANP is more advanced than AHP because it includes the relationships between elements within the structure.

Network structures are integral to ANP. In the network structure, a node is a cluster of associated elements; the lines within the structure suggest interaction between clusters, while the inner loop represents dependencies between the elements within a cluster. There are two types of influences/dependencies between elements considered in ANP: internal and external. Internal influences are dependencies of one element on another element within a cluster. External influences are the effect that elements from one cluster have on the elements of another [16]. The advantage of ANP is that it is able to determine the priority of clusters and their elements. In addition to considering the interdependencies of elements it also takes into account the independent elements themselves [41].

With ANP, interactions between elements are established by applying a supermatrix. Within MCDM a supermatrix includes three types of relations [17]: (1) independence from successive criteria/sub-criteria, (2) interdependence between criteria/sub-criteria and (3) interdependence between the levels of criteria and sub-criteria. ANP can be implemented in six steps as presented in Fig. 1:

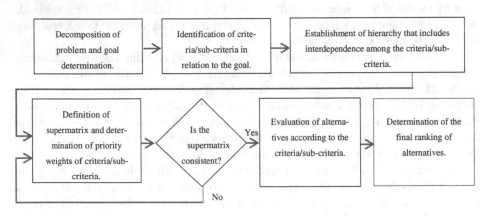

Fig. 1. The steps of ANP

3.2 Decision Making Trial and Evaluation Laboratory - DEMATEL

DEMATEL was developed between 1972 and 1979 by the Science and Human Affairs Program of the Battelle Memorial Institute of Geneva [11]. The goal of DEMATEL is to convert the causal relationships between elements from a complex system to an understandable structural model [21]. DEMATEL is helpful in visualizing the structure of complex causal relationships between evaluation criteria through the use of matrices or digraphs [4].

This method involves two groups - causal and effect. The causal group affects the effect group and thus are determined the weights of criteria [7]. This technique allows decision-makers better comprehension of the structural relationship between elements of a system [45]. This method is applied to analyze and outline the relationship of cause and effect between the evaluation criteria [43] or to reveal interrelationship between factors [21]. Phases of the DEMATEL method can be presented as in the following Fig. 2:

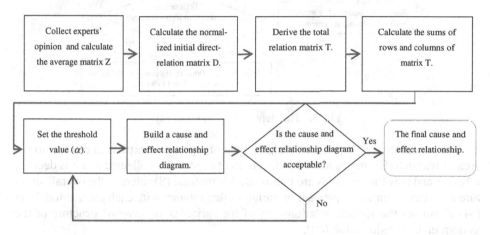

Fig. 2. The steps of DEMATEL (Source: [37])

DEMATEL analyzes the structure of components within each criterion, as well as the intensity of direct and indirect relationships between the defined components, causal relationships, and the strength of influence [20]. Structural matrices and causal diagrams are used to present the causal relationships and levels of impact between criteria in a complex system.

3.3 Interpretive Structural Modelling - ISM

ISM is proposed by Warfield [42] for the analysis of complex social and economic systems. ISM presents a computer-assisted learning process that allows individuals or groups to develop a map of the intricate relationships between the different elements that are involved in a complex situation [3]. The basis of this technique is to use the practical knowledge of experts to decompose a complicated system into several sub-systems, i.e. to create a structural model which consists of several levels (Fig. 3).

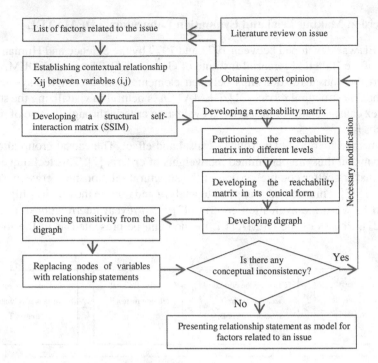

Fig. 3. The steps of ISM (Source: [3])

The interpretive (I) represents the judgment of a group of experts in relation to the area of research. Estimates made by the group of experts are collected and it is decided whether and how the variables are interrelated. Structural (S) refers to the overall structure extracted from a complex set of variables that interact with each other. Modelling (M) illustrates the specific relationships of the variables and overall structure of the system under consideration [12].

3.4   Fuzzy Measures and the Choquet Integral - CI

Fuzzy measures are set functions with monotonicity, which aren't necessarily additive [34]. In other words, they are an extension of a measure in the sense that the additivity of the measure is replaced with a weaker condition, monotonicity. Sugeno [36] also suggests the fuzzy integral, which is an integral with respect to fuzzy measures. The CI is suggested within the fuzzy measure community by Murofushi & Sugeno [25], while the fuzzy integral was put forward by Choquet [6] and encompasses interactions between k out of n criteria of the problem, which is called the k-additivity property.

The CI is an approach that is closely related to fuzzy measures. Fuzzy integrals are used to present the interactions between criteria. They allow simple translation of a decision-maker's requests into coefficients of the fuzzy measure. The idea is that super-additivity of the fuzzy measure implies synergy between the criteria, and subadditivity implies redundancy [14]. Most real applications of fuzzy measures deal with MCDM problems, where fuzzy measures are defined on the finite set of criteria, and model the relative importance of criteria as well as their interaction [13]. Furthermore, Grabish & Roubens [15] have proposed an axiomatic basis for the interaction index, giving a consistent basis for dealing with the notion of interaction.

3.5 Interpolative Boolean Algebra - IBA

Since conventional fuzzy set theory does not satisfy all Boolean axioms and laws, methods for the consistent realization of fuzzy logic have been developed. Consistent generalization of fuzzy logic is enabled by using Interpolative Boolean Algebra (IBA), as proposed by Radojevic [29]. IBA is a real valued, and/or [0,1] value realization of Boolean algebra [8]. This approach includes all logical functions, interpolative operators and generalized product operators [27]. Under IBA, all Boolean axioms and theorems apply [28].

IBA has a finite number of elements and it is atomic algebra. IBA clearly separates the structure and value of the elements of Boolean algebra. It consists of two levels: (a) symbolic and (b) valued [27]. At the symbolic level, one of the basic concepts is the structure of elements in IBA. The principle of structural functionality indicates that the structure of any combined element in IBA can be directly calculated based on the structure of its components. The valued level is a concrete symbolic level in terms of value. An element from the symbolic level preserves all its characteristics at the value level, as described by Boolean axioms and laws [28].

IBA is technically based on the generalized Boolean polynomial – GBP [29]. GBP is a polynomial of which the variables are elements of Boolean algebra, and thus it allows for the processing of the corresponding element of Boolean algebra into the value of the real interval [0,1] using operators such as classical ($+$), classical ($-$) and generalized product ($\otimes$) [27].

4 Discussion

In addition to the presented methods, it is necessary to mention the more recent approaches which include interactions between criteria in their analysis. Among them are the following:

MUSA (Multi-Criteria Satisfaction Analysis), proposed by Angiella et al. [2], represents a preference disaggregation approach that is based on the principle of ordinal regression analysis. MUSA finds an additive utility function that represents the level of satisfaction of users based on their preferences. By using this approach, users determine the comprehensive satisfaction level for each product/service, but the marginal satisfaction level for each criteria of the decision-making is also determined.

UTA (Utilites Additives) includes robust ordinal regression, and is proposed by Jacquet-Lagreze & Siskos [18]. UTA belongs to the utility/value function category of MCDM approaches. Slowinski et al. [33] have proposed an extended version of the UTA method for the assessment of strong or weak outranking relations and the problem of multi-criteria ranking. This method takes into account all compatible value functions at the stage of ranking.

Figueira et al. [10] have proposed an improved version of ELECTRE (Elimination and Choice Expressing the Reality) which involves the analysis of interactions between criteria. Specifically, it expands the concept of concordance and discusses three types of interactions designated as mutual strengthening, mutual weakening, and antagonistic.

From the techniques presented, it can be concluded that MCDM methods which include interactions between criteria use different tools for their presentation, for example: ANP - hierarchy structures and supermatrices; DEMATEL - structural matrices and cause/effect diagrams; ISM - structural self-interaction matrices; fuzzy measures and CI - fuzzy measures and integrals; and IBA - logical function Boolean operators, GBP and LA. The following table (Table 1) presents the basic advantages and disadvantages of the stated methods.

Table 1. Advantages and disadvantages of presented MCDM methods

Method	ANP	DEMATEL	ISM	Fuzzy measures and integrals	IBA	MUSA	UTA	ELECTRE
Advantages of methods	Takes into account dependent and independent criteria	Determines the direct and indirect relationships between criteria	Allows judging of differences between elements and understanding of what criteria are based on	Presents positive and negative interactions between criteria	Provides structural transformation rather than introducing numerical values	Considers qualitative form of customers' judgments and preferences	Complete ranking using one compatible value function	Takes into account three types of interaction: mutual strength, mutual weakness and antagonistic
Disadvantages of methods	Is unable to single out an element and identify its strengths and weaknesses	Mechanism which allows the integration of indirect/direct relation is unclear	Cannot consider a number of criteria and is not statistically validated	Evaluates all criteria on one scale	Decision-makers cannot always adequately set logical functions	Cannot present positive and negative synergies	Unrealistic hypothesis concerning preferential interdependence between criteria	Is not adequate for representing a large number of interactions between criteria

5   Conclusion

Many MCDM methods are effectively used for solving a large number of decision-making problems in different fields of research. However, some of the most famous MCDM methods such as AHP and TOPSIS only allow the establishment of linear relationships between elements of decision making. Considering the relationship between criteria/sub-criteria in linear form only, ignoring any mutual interaction or interdependence, has proven not to be adequate in practice.

Academics, therefore, began to develop holistic approaches involving interrelationships between the elements of decision-making. Some of the approaches discussed in this paper are ANP, DEMATEL, ISM, fuzzy measures and integrals, and IBA. Analysis of the presented methods demonstrates that they give more reliable results by viewing the aggregation of criteria as a non-linear structure where elements are interrelated. Most MCDM methods, in structuring complex decision-making models, viewed criteria as independent elements. In many complex real world decision problems, however, there are certain relationships and interdependencies between the criteria. Moreover, the value of the criteria for an appropriate action can be influenced by a number of factors that are external to the decision system and cannot be controlled by the decision-maker.

In this paper we would like to highlight the importance of modelling the interactions betwen criteria in decision-making, because otherwise it may lead to making bad decisions according to false assumptions of linearity and independence. As can be concluded from the presented research, all recently proposed MCDM methods are dealing with the inclusion of interactions between decision-making criteria. Therefore, with further research, we aim to identify all of the newly developed approaches, perform a comparison with existing methods, and determine which of these routes is most appropriate to establish relations between decision-making criteria.

References

1. Ananda, J., Herath, G.: A critical review of multi-criteria decision making methods with special reference to forest management and planning. Ecol. Econ. **69**, 2535–2548 (2009)
2. Angilella, S., Corrente, S., Greco, S., Slowinski, R.: MUSA-INT: multicriteria customer satisfaction analysis with interacting criteria. Omega **42**, 189–200 (2014)
3. Attri, R., Dev, N., Sharma, V.: Interpretative Structural Modelling (ISM) approach: an overview. Res. J. Manage. Sci. **2**(2), 3–8 (2013)
4. Chai, J., Liu, J.N.K., Ngai, E.W.T.: Application of decision-making techniques in supplier selection: a systematic review of literature. Expert Syst. Appl. **40**(10), 3872–3885 (2013)
5. Chen, F.H., Hsu, T.S., Tzeng, G.H.: A balanced scorecard approach to establish a performance evaluation and relationship model for hot spring hotels based on hybrid MCDM model combining DEMATEL and ANP. Int. J. Hosp. Manage. **30**, 908–932 (2011)
6. Choquet, G.: Theory of capacities. Annales de L'Institute Fourier. **5**, 131–296 (1954)
7. Dalalah, D., Hayajneh, M., Batieha, F.: A fuzzy multi-criteria decision making model for supplier selection. Expert Syst. Appl. **38**(7), 8384–8391 (2011)
8. Dragovic, I., Turajlic, N., Radojevic, D., Petrovic, B.: Combining Boolean consistent fuzzy logic and AHP illustrated on the web service selection problem. Int. J. Comput. Intell. Syst. **7**(1), 84–93 (2013)
9. Falatoonitoosi, E., Leman, Z., Sorooshian, S., Salimi, M.: Decision-making trial and evaluation laboratory. Res. J. Appl. Sci. Eng. Technol. **5**, 3476–3480 (2013)
10. Figueira, J.R., Mousseau, V., Roy, B.: ELECTRE methods. In: Figueira, J.R., Greco, S., Ehrgott, M. (eds.) Multiple Criteria Decision Analysis: The State of the Art Survey, pp. 133–162. Springer Science, Business Media Inc., New York (2005)
11. Fontela, E., Gabus, A.: The DEMATEL observer, DEMATEL 1976 report. Battelle Geneva Research Center, Switzerland Geneva (1976)
12. George, J.P., Pramod, V.R.: An interpretive structural model (ISM) analysis approach in steel re rolling mills (SRRMs). Int. J. Res. Eng. Technol. **2**(4), 161–174 (2014)
13. Grabisch, M.: A survey of applications of fuzzy measures and integrals. In: 5th International Fuzzy Systems Association Conference, Seoul, Korea (1993)

14. Grabisch, M.: Fuzzy integral in multicriteria decision making. Fuzzy Sets Syst. **69**, 279–298 (1995)
15. Grabish, M., Roubens, M.: Application of the Choquet integral in multicriteria decision making. In: Grabisch, M., Murofushi, T., Sugeno, M. (eds.) Fuzzy Measures and Integrals - Theory and Applications, pp. 348–374. Physica Verlag, Heidelberg (2000)
16. Gürbüz, T., Albayrak, Y.E.: An engineering approach to human resources performance evaluation: hybrid MCDM application with interactions. Appl. Soft Comput. **21**, 365–375 (2014)
17. Gürbüz, T., Alptekin, S.E., Alptekin, G.I.: A hybrid MCDM methodology for ERP selection problem with interacting criteria. Decis. Support Syst. **54**, 206–214 (2012)
18. Jacquet-Lagrèze, E., Siskos, Y.: Assessing a set of additive utility functions for multicriteria decision making: the UTA method. Eur. J. Oper. Res. **10**, 151–164 (1982)
19. Keeney, R.L., Raiffa, H.: Decisions with Multiple Objectives: Preferences and Value Tradeoffs. Wiley, New York (1976)
20. Lee, W.-S., Huang, A.Y., Chang, Y.Y., Cheng, C.M.: Analysis of decision making factors for equity investment by DEMATEL and analytic network process. Expert Syst. Appl. **38**(7), 8375–8383 (2011)
21. Lin, C.L., Tzeng, G.H.: A value-created system of science (technology) park by using DEMETEL. Expert Syst. Appl. **36**, 9683–9697 (2009)
22. Mandić, K., Delibašić, B., Radojević, D.: Supplier selection using Interpolative Boolean algebra and TOPSIS method. In: Group Decision and Negotiation Conference 2014, Proceedings of the Joint International Conference of the INFORMS GDN Section and the EURO Working Group on DSS, Toulouse University, France, pp. 134–142 (2014)
23. Meade, L., Sakris, J.: Analyzing organizational project alternatives for the agile manufacturing process: an analytical network approach. Int. J. Prod. Res. **37**(2), 241–261 (1999)
24. Mehregan, M.R., Hashemi, S.H., Karimi, A., Merikhi, B.: Analysis of interactions among sustainability supplier selection criteria using ISM and fuzzy DEMATEL. Int. J. Appl. Decis. Sci. **7**(3), 270–294 (2014)
25. Murofushi, T., Soneda, S.: Techniques for reading fuzzy measures (III): interaction index. In: 9th Fuzzy System Symposium, Sapporo, Japan, pp. 693–696 (1993)
26. Nguyen, H.T., Dawal, S.Z.M., Nukman, Y., Aoyama, H.: A hybrid approach for fuzzy multi-criteria decision making in machine tool selection with consideration of the interactions of attributes. Expert Syst. Appl. **41**, 3078–3090 (2014)
27. Radojevic, D.: Interpolative realization of Boolean algebra as consistent frame for gradation and/or fuzziness. In: Nikravesh, M., Kacprzyk, J., Zadeh, L.A. (eds.) Forging New Frontiers: Fuzzy Pioneers II. STUDFUZZ, vol. 218, pp. 295–318. Springer, Heidelberg (2008)
28. Radojevic, D.: Interpolative realization of Boolean algebra. In: Proceedings of the NEUREL 2006, The 8th Neural Network Applications in Electrical Engineering, pp. 201–206 (2006)
29. Radojevic, D.: Logical measure of continual logical function. In: 8th International Conference IPMU – Information Processing and Management of Uncertainty in Knowledge-based Systems, Madrid, pp. 574–578 (2000)
30. Saaty, T.L.: Decision making in Complex Environments, The Analytical Hierarchy Process for Decision Making with Dependence and Dependence and Feedback. RWS Publications, USA (1996)
31. Saaty, T.L.: The Analytic Hierarchy Process. McGraw-Hill, New York (1980)
32. Sage, A.P.: Interpretive Structural Modeling: Methodology for Large Scale Systems. McGraw-Hill, New York (1977)

33. Slowinski, R., Greco, S., Matarazzo, B.: Rough sets in decision making. In: Meyers, R.A. (ed.) Encyclopedia of Complexity and Systems Science, pp. 7753–7786. Springer, New York (2009)
34. Sugeno, M., Murofushi, T.: Pseudo-additive measures and integrals. J. Math. Anal. Appl. **122**, 197–222 (1987)
35. Sugeno, M.: Fuzzy measures and fuzzy integrals – a survey. In: Gupta, M., Saridis, G., Gaines, B. (eds.) Fuzzy Automata and Decision Processes, pp. 89–102. North-Holland, Amsterdam (1977)
36. Sugeno, M.: Theory of fuzzy integrals and its applications. Ph.D. thesis, Tokyo Institute of Technology (1974)
37. Sumrit, D., Anuntavoranich, P.: Using DEMATEL method to analyzes the casual relations on technology innovation capability evaluation factors in thai technology-based firms. Int. Trans. J. Eng. Manage. Appl. Sci. Technol. **4**(2), 81–103 (2013)
38. Tadić, S., Zečević, S., Krstić, M.: A novel hybrid MCDM model based on fuzzy DEMATEL, fuzzy ANP and fuzzy VIKOR for city logistics concept selection. Expert Syst. Appl. **41**, 8112–8128 (2014)
39. Toloie-Eshlaghy, A., Homayonfar, M.: MCDM methodologies and applications: a literature review from 1999 to 2009. Res. J. Int. Stud. **21**, 86–137 (2011)
40. Umm-e-Habiba, Asghar, S.: A survey on multi-criteria decision making approaches. In: International Conference on Emerging Technologies, ICET, pp. 321–325 (2009)
41. Velasquez, M., Hester, T.P.: An analysis of multi-criteria decision making methods. Int. J. Oper. Res. **10**(2), 56–66 (2013)
42. Warfield, J.W.: Developing interconnected matrices in structural modelling. IEEE Trans. Syst. Men Cybern. **4**(1), 51–81 (1974)
43. Yang, Y.P., Shieh, H.M., Leu, J.D., Tzeng, G.H.: A novel hybrid MCDM model combined with DEMATEL and ANP with applications. Int. J. Oper. Res. **5**(3), 160–168 (2008)
44. Zadeh, A.L.: Fuzzy sets. Inf. Control **8**(3), 338–353 (1965)
45. Zhou, Q., Huang, W., Zhang, Y.: Identifying critical success factors in emergency management using a fuzzy dematel method. Saf. Sci. **49**, 243–252 (2011)
46. Zopounidis, C., Doumpos, M.: Multicriteria classification and sorting methods: a literature review. Eur. J. Oper. Res. **138**, 229–246 (2002)

Author Index

Athanasiadis, Ioannis 73

Bobar, Vjekoslav 98
Bohanec, Marko 46

Caroleo, Brunella 34
Chen, Huilan 22
Costa, Ana Paula Cabral Seixas 10
Cunha, Annielli 85

de Carvalho, Victor Diogho Heuer 10
Delibašić, Boris 46, 98
di Donato, Francesca 61

Hao, Yuqiuge 22

Liu, Shaofeng 22

Mandic, Ksenija 98
Morais, Danielle 85

Neaga, Irina 22
Nieddu, Luciano 61

Osella, Michele 34

Papathanasiou, Jason 73
Ploskas, Nikolaos 73
Poleto, Thiago 10
Power, Daniel J. 1

Samaras, Nikolaos 73

Tosatto, Andrea 34

Xu, Lai 22

Printed in the United States
By Bookmasters